Sex, Drugs and DNA
Science's taboos confronted

Michael Stebbins

Macmillan
London   New York   Melbourne   Hong Kong

First published in hardback 2006
First published in paperback 2007 by
Macmillan
Houndmills, Basingstoke, Hampshire RG21 6XS and
175 Fifth Avenue, New York, N. Y. 10010
Companies and representatives throughout the world

ISBN-13: 978–1–4039–9342–7 (hardback)
ISBN-10: 1–4039–9342–4 (hardback)
ISBN-13: 978–0–230–52112–4 (paperback)
ISBN-10: 0–230–52112–6 (paperback)

This book is printed on paper suitable for recycling and made from fully
managed and sustained forest sources.

A catalogue record for this book is available from the British Library.

A catalog record for this book is available from the Library of Congress.

10	9	8	7	6	5	4	3	2	1
16	15	14	13	12	11	10	09	08	07

Printed and bound in China

Contents

0

it's enough to make you ...

Let me make one thing absolutely clear: this is not a scholarly tome. There are no references, interviews or balanced arguments. It is, in essence, an experiment. You see, the public often finds science books boring, and frankly – so do I. They can be very pretentious. But science is anything but pretentious though. It is really a humbling endeavor. The decision to write this book did not come out of any overwhelming need to document moments of discovery, or the history of science, or to make balanced scientific arguments. It came from my perception of a need to address the litany of bullshit and lies spewed at the public. This sort of no-holds-barred writing is virtually never used in science.

I have stripped out the jargon and tried to deal with science and health issues as though I was haranguing you at a party. Hopefully, I'll give you some laughs as well. Since most people are concerned with how science is going to affect their lives, I also discuss healthcare – drugs in particular – extensively. Got to feed the narcissists! I point the finger when I think it needs pointing and try to give the straight story.

The strange thing about talking to non-scientists about science is that you quickly notice that some of the smartest, most thoughtful and intellectually curious people have a terrible understanding of it. Most students leave high school poorly equipped to manipulate even the most basic concepts. Nonetheless, I remain hopeful that we can increase science literacy and intelligent discourse in the US. What can I say? I'm an optimist.

As advances in science have became more commonly reported in the media over the last ten years, public questions and concerns about science have become more common. But scientists have failed to communicate science to the public in a language they can understand. This is because scientists speak in a kind of Klingon language of their own. They just don't instinctively shift to more colloquial explanations of their work on the rare occasion that they address non-experts.

Growing attacks on the integrity of science by politicians and religious leaders and the consistent lack of emphasis on science and health education in the public school system has compounded the situation. These people have done everything they can to interfere with scientific advance, but have really only succeeded in giving much of the public the impression that scientists are a bunch of left wing elitists with a social agenda. This is rubbish, although the attacks from right wing conservatives have driven the great majority of the scientific community to the political left.

I have deliberately glanced over many ideas and have most certainly not given a thorough accounting of each of the subjects. To do so would require an entire book for each subject. Mainly I try to address the stuff that pisses me off the most.

I resent many of the people that make this book necessary. I am talking about the miscreants and bent backs that deliberately try to confuse the public about science and have attacked scientists to move social or political agendas that go against the best information we have. We are all a little dumber for having put up with their garbage. We should all take every opportunity to point out how sick and perverted they have become.

The Department of Education has done a horrible job in fostering sound science education in the US. I apologize to the students who have not been inspired through solid science education because of their incompetence. I am actually apologizing for your parents. They voted in a bunch of idiots who have not done their part to ensure that the public school system is serving you best. I had two teachers when I was growing up, Barry Stow and Keith Lauber, who made all the difference in the world. Neither of them were science teachers, but these two men dedicated their careers to making education fun and sparking students to think critically based on facts, so they might as well have been. We could use a lot more of that. Tom Brokaw describes your grandparents, who fought through World War II and the Great Depression as the 'Greatest Generation.' Let me assure you that the baby boomer parents, who fought through white wine spritzers in the 1980s and the HOV lane in their SUVs will go down as the worst generation. They have failed us all.

Often I have made no attempt to present a balanced argument, especially where I don't see balance. When there is truly significant scientific uncertainty about a subject, I have done my best to explain it. Where the overwhelming information points in one direction, I see little reason to honestly consider fringe views and in some instances have gone out of my way to marginalize them. This is completely counter to the way science is done, but it is important to make clear arguments to a people deluged with false information. I have also suggested solutions to many of the problems that plague national science policy and healthcare. They are meant to be suggestions and they do not represent the view of the scientific community. Scientists don't work that way. They just don't put ideas forward as a group, and this has been both their strength and their weakness.

Since the first printing of this book, a lot has happened. So I have tried to update all of the relevant information to reflect the current state of science and the world. I have heard from a great number of people who read the first printing. Most were extremely complimentary, others were mad that they had been duped by their leaders and a select few were mad at me because they perceived me as being anti-American. That was to be expected. What was unexpected were the letters from people who drew hope from it. Several had loved ones who were ill. Others were excited about science for the first time in their lives or the first time in a while. Those letters were flattering and inspiring. I hope I can live up to their gestures of kindness. I feel a deep humbling gratitude to the people who took the time to tell me how much it meant to them. To those people...NO: thank *you*.

I know some people are going to screw this up, but I'll say it anyway – the opinions expressed in this book are my own and do not in any way represent the view of any people or organizations I am affiliated with, including anyone mentioned in the book. I do not speak for any group here.

I could not have written this book without the help of a large number of people who have supported me throughout my career, including some of the nontraditional twists it has taken. Many people who have been instrumental in my career would prefer not to be mentioned here because of some of the criticisms and strong opinions that I put forward.

First, I thank my family. While demented and annoying, they are my rock. Writing this book would not have been nearly as enjoyable if it weren't for the advice and support of my friends at Cold Spring Harbor Laboratory, SUNY Stony Brook, the Nature Publishing Group, the National Human Genome Research Institute, the American Society for Human Genetics,

ScienCentral News, the United States Senate and House of Representatives, the National Institutes of Health, the Federation of American Scientists, Scientists and Engineers for America, Johns Hopkins University, Harvard University, Columbia University Rutgers University, Massachusetts Institute of Technology, Robert Wood Johnson Medical School, Oxford University, Columbia University, Yale University, Rockefeller University and UNLV, many of whom were a tremendous help with the content.

I'd also like to thank Matt Sesow, an extraordinary artist and dear friend, for helping me to rescue my computer when I was in a real pinch.

I could not have written this book without inspiration from The Ramones, Howard Stern, The Daily Show, Richard Feynman, Michael Moore, and a host of irreverent artists, scientists, musicians and professional troublemakers that I call friends.

As I said, much of this book was born during conversations with friends who have served as guinea pigs for much of the content. I would like to thank Kim Weinreich in particular for keeping me mostly sane. She is a remarkable person. All of the cartoons were drawn by Dr Sean Taverna, who, in addition to being an extraordinary cartoonist, is also a very talented scientist. My dear friend Daphne Youree kindly agreed to photograph me and has a knack for making me look more handsome than I am. If it were not for Ian Kingston, you would have had the opportunity to see how badly I can butcher the English language. This book would not have happened if it were not for my editor Sara Abdulla, who was convinced early on that I could write something interesting.

Finally, I would like to dedicate this book to my mother, Kathy, who taught me that fighting tyranny is better than living with it. For more information and my blog please visit www.sexdrugsanddna.com/blog.

1

a scientist's life

Forget everything you think you know about science and scientists. It's probably a bunch of crap. Before we dive into the controversies that surround science, I am going to give you a peek into the making of a typical modern biologist. This is important because without some context of what scientists do, it is hard to understand how they must feel about the current denial and misuse of science to push social and political agendas. I will focus shamelessly on the US for one simple reason: the US is a scientific goliath. It produces about the same number of scientific papers per year as all of Europe combined. Some 63% of the top 1% of all research papers in the biological sciences, measured by citation, are produced in the US. Simply put, the US produces more science that is cited more often by other scientists than the rest of the world. There is no significant biomedical research being published in any language other than English. Good science is being done elsewhere; there is just more of it being done in the States.

In the beginning

A budding biological scientist is faced with a major decision at some point in her undergraduate years: do I want to be a physician or do I go into research? Looking around the classroom and seeing the spreading plague of premeds is an unsettling experience for those who are born to question. Our future scientist finds herself liking the rigor of physics classes and organic chemistry. She loathes the grade whores who crowd professors at the end of class. She might even be tempted to do a little grade whoring herself, but feels uncomfortable with it. But very few labs will take undergraduate freshmen and sophomores in. They are seen as being too young, inexperienced and unreliable. Meantime, they must endure their colleagues' shallowness and eke out an education in classes designed for premeds.

Scientists are born in the lab. It is there that an undergraduate is expected to learn quickly and accomplish something.

Scientists do not get extra credit for having undergrads in their labs and no one gives them significant grant money to spend their time with undergrads. It is the place where the coddling of high school and the undergraduate classroom disappears. Students who are serious about learning how to do science will succeed in this environment. Those who are just looking to pad their resumes with another line and get themselves a letter of recommendation for medical school will rarely contribute anything significant and typically don't last more than a semester.

There is no golden path of social and economic avarice laid out for scientists. They get the bug for research out of curiosity and a drive to do something that no one else has ever done. While somewhat misguided, the thrill of your first experimental results is like crack to a future scientist. Studying and jockeying for grades will be shed for a new passion for experimentation.

Our student has now decided that she has no desire to stitch wounds, prescribe blood pressure medication or provide strippers with bust enhancement surgery. Instead, she finds herself spending more and more time in the lab and decides that getting a PhD in biomedical sciences is her path in life. Instead of six figures' worth of debt, she can get a free education ... actually, tuition is free and she will get *paid* to go to graduate school. No annoying patients and no old people complaining about their gout, and you get pocket money instead of debt – where do you sign up? The promises of graduate school sound like an ad for the military, and might be nearly as divisive. 'Pay' is really a bit of an exaggeration for what students are actually getting for their effort. In reality, they will earn less than minimum wage per hour. Their salary is often not enough to pay the bills, requiring student loans or money borrowed from family to make up the difference.

At some institutions, graduate students are the primary source of research results and thus bring in the bulk of the money. Research results equal grants. At many universities and institutions, a great deal of the university rides on the money brought in by biomedical research grants. Few realize that the philosophy, history and other humanities departments at universities don't bring in a lot of dough. When a biomedical researcher brings in a grant, a percentage of the grant goes directly to the university for 'overhead' costs. This can be more than half of the grant, which pretty much sucks. What that means is that science research floats the non-profitable departments and in many cases is the only reason they can persist. Professors in these departments are often paid as much as biomedical researchers – sometimes more. A close look at many universities reveals that the highest-paid faculty staff are often the ones who have been there the longest and brought in the least money. There is no greater waste of time, money and space than a professor emeritus at a university who is taking a paycheck but not contributing to the department he is in. These professors don't teach much or bring in money, but can draw big salaries. Perhaps they will write a book of the collected wisdom of a career in academia. That's certainly valuable right? Not usually. When these books are written, the income from them goes right into the professors' pockets, so it doesn't subsidize their salary, and the books usually fall somewhere between stuffy and sucky (with a handful of notable exceptions). Advice to university administrators – burn the dead wood.

So, our student applies to grad school and is off to a life as an academic researcher. All is well – her parents can brag that she is curing cancer and she will lead a comfortable life ... Right? *Wrong*! The hurdles are now lined up and chances are she won't make it to the end of the race. First is graduate school. Getting into a good graduate program is key, because a PhD from Podunk U. is virtually useless. To get into a good graduate program, a student will obviously have to have a combination of good grades and admissions test scores. But she will also have to have significant research experience *and* will have to have actually proven that she is competent in the lab. The best way to show an admittance committee that you are competent enough to succeed is to have a recommendation by a respected researcher. Most grad school applicants don't have authorship on research papers, so a written description of the project that they worked on and the recommendation is what they have to rely on. If one does undergraduate research in the lab of a big-time scientist, then more weight will be given to the recommendation. This is not by design – it is just human nature. If those sitting on the committee know the recommender, the recommendation will be trusted and the student will be given a close look. This again shows that the undergraduate institution is key: Dr Big-Time Scientist is unlikely to be working at Podunk's sister school out in rural Nebraska.

Let the games begin

Once in graduate school, the new student will settle in for a year of what is best described as hazing. A typical grad student will have to split her time taking classes, studying for tests, teaching undergraduate classes, and doing research as a rotation student for the first year. It should be noted that this is likely to be the only teaching experience she will ever have before she is actually required to organize and teach lectures to large classrooms of students as an assistant professor. It is completely normal for a first-year student to leave before 8 in the morning and not come home until 11 or 12 at night. As a rotation student (or 'roton'), the student spends several months working in a lab on a project. The purpose of the rotation is similar to that of medical school rotations. They are an opportunity for the student to decide if the particular research projects going on in the lab are best suited for them and whether they will be able to work with the particular personalities in that lab – most importantly the primary investigator, or PI, who is running the lab. At the same time, the rotation serves as a long job interview for the student. The PI and the people in the lab are judging not only how well the student does science, but whether the person's personality jibes well with that of the lab. After all, the student is likely to spend 5–6 years or more working in that lab – the longest and biggest career commitment that the student has ever made.

Our grad student teaches a laboratory class associated with a larger lecture class, which is in turn taught by a professor. She serves as a teaching assistant for that class, running review sessions, giving practical lectures, grading tests and homework and holding office hours during which premed students will try to suck out information on what will be on the test. The very students that she shared classes with just three months earlier as an undergraduate are now coming to her for information.

During this first year she will also have to take classes. It is pretty normal for grad programs to require their students to get a minimum of a B in any class they take. Sometimes these classes are taken with medical students, who at this point have gone through a welcome transition where they are more comfortable that they are going to become doctors. But our itinerant roton is torn. She feels some responsibility to her grades, but is pressured to spend as much time as she can in her rotation labs doing research, learning new techniques and impressing her rotation PIs. She also feels a responsibility to her own students, but more often than not, if someone is going to suffer, it is the undergrads, who will have little to say about what happens to the roton in the coming months.

The devotion of scientists is rarely in question. What is in question is what they are devoted to. Scientists will suffer at the hands of their own passion and drive for their entire career. Relationships suffer in particular, and the first year of grad school should be an eye opener to budding scientists that something is not right. Relationships come and go in everyone's life, but what is interesting amongst scientists is that devotion to the craft of being a scientist is a primary reason for relationships to crumble. If there is anything wrong with your relationship when you enter grad school, the pressure and long hours will bring it right to the surface. Those students who are not in relationships will often hook up with their fellow students, who can at least empathize with the new lifestyle that they have entered. Still others, despite being expressly forbidden from doing so, wind up dating their undergraduate students. This is actually more common than people would suspect, and there is little that grad schools can do to prevent it. Every graduate student knows a colleague who has dated one of their students – or worse, a professor. This too is strictly forbidden at all universities, but it happens all the time. Some universities have made concessions to situations where a student and faculty member can see each other as long as the faculty member has no say over the student's fate. Here's an interesting example.

In 2003, Arnie Levine, the president of Rockefeller University, was dismissed from his position and lambasted in the *New York Times* for dating a graduate student at the University. Dr Levine is a well-respected member of the research community whose life's work was rewarded with a prestigious position at the University in 2000. What was his crime? He showed up in a public place with his girlfriend, who, by the nature of his position, was someone he should not have had an intimate relationship with. Unofficially, there were a series of events that had led up to his dismissal, but what the *New York Times* failed to report was that this sort of relationship happens at universities *all the time*. They also missed the fact that these relationships are tolerated as long as faculty members and teaching assistants can maintain a reasonable level of plausible denial. Many research institutes have gone so far as to settle harassment cases out of court to save face and preserve the careers of professors who have had inappropriate relationships. In the absence of complaints, it is more typical for universities to turn a blind eye to these indiscretions. Dr Levine removed the element of plausible deniability, thus risking embarrassment to the university, and lost his job because of it.

Once our roton has made it through her first year without engaging in an inappropriate relationship, has found a lab and is starting a project, she will settle into her new life as an indentured servant. Different graduate schools have different policies regarding what is required of a student to receive a PhD. It is fairly common for a student to work on a project for a year and then submit a proposal for a thesis project. At this point the student will put together a committee of professors from the university and perhaps one from an outside university. In effect, the student has a primary advisor (the primary investigator for the lab she is working in) and a committee of other professors who must approve both the student's proposal and ultimately her final thesis. The committee serves two capacities. First, it provides advice to the student on a range of topics, such as experimental design and how to deal with an unreasonable thesis advisor. Second, the committee members serve as advocates for the student so that the advisor cannot hold the student hostage in graduate school for too long or ask her to do anything that they perceive as unreasonable. Sounds like a good system right? Bzzz! Sorry, thank you for playing. We have some nice parting gifts for you.

The committee members do indeed serve as effective advisors for experimental design and will generally give the student good advice on achieving success in the research project. However, when a student has a personal or professional problem with their advisor, the committee is usually powerless to do anything about it. Strike that – the committee isn't actually powerless. It's impotent. The committee would like to do something about an advisor who is being unreasonable, but many times cannot do anything because if they do dare strong-arm the primary investigator, they risk putting themselves in a bad position with a fellow faculty member. This is particularly true of committee members who happen to be junior faculty looking to get tenure in the coming years. There are of course exceptions, but this is usually the case. So our student, in many cases, will feel like she is out there on her own. The system is set up so that students go through a stage where they feel helpless and under-appreciated, and see no end in sight.

There is a honeymoon period in which our young scientist and her new advisor will work well together. Everything is new and potential abounds. It is in the best interests of advisors to get as much as they can from students once they are productive in the lab. This translates into research papers. Most schools require their students to have at least one research publication before they graduate; some require two or three. This is so that the students can be competitive in getting grants as post-doctoral fellows and to make sure that the lab and institution are getting something out of the students they train. What this does in effect is ensure that most students take projects that are likely to succeed and in most cases will be of little consequence to their field. In essence, they do boring work to ensure success.

The situation is quite simple. The currency of biomedical research is the research paper. There is an adage in science: publish or perish. There could be no truer words. Most researchers' salaries are paid from research grants. Some are paid by foundation grants or venture capital money dumped into the little biotech company they have on the side, but the overwhelming majority are paid with taxpayer dollars through the National Institutes of Health, the National Science Foundation or other government-supervised trough. The funding system is set up so that only about 7–12% of research grants submitted are actually funded by the NIH, and this has become tougher over the last couple of years. Writing these grant applications is a time-consuming, frustrating and ultimately unrewarding process. I say unrewarding, because the scientist gets very little scientifically from the process – except money of course. Assuming, that is, that they successfully navigate the process. It can take weeks of solid work to put together an NIH grant, weeks that could be spent more productively if the process wasn't so competitive and if the grants themselves weren't so bureaucratically complicated and unbelievably detailed – 40 pages of written material is not uncommon.

What scientists are actually doing is justifying their research over and over again through the course of their career. They repeatedly go through this process of peer review. It is an unbelievable system of checks and balances that is supposed to keep people on track and make sure that grant money is not being spent frivolously. Researchers can't just go out willy-nilly and do whatever they like. The supposed freedom of academics that people opposed to research funding tout as a reason to reduce NIH spending is a complete fallacy, and no one has bought into it more than the researchers themselves.

Our student works out her project with her advisor, gets some preliminary results that indicate that the project will work and she meets with her committee. They evaluate her proposal and let her move on. If she had done poorly and her project had looked like it would bomb or she had displayed a lack of knowledge about her project, she would have failed her proposal and would have been kicked out of school. Getting kicked out of graduate school would seem like a really bad thing. It isn't. If our student makes it to her thesis proposal and doesn't know what she is doing, or fails classes or exams multiple times, then it is best if she gets out and does something else with her life. There's always medical school. In the end, whether a student is kicked out is really at the discretion of the advisor. I have never heard of a single case where a student was kicked out after failing the proposal if the advisor still supported the student's efforts. Here is where the student's ability is really not the issue so much as the advisor's willingness to support her.

Let's jump ahead. Our student passes her proposal and has immersed herself in her project. She is now dating a fellow grad student; wisely, he doesn't even work on the same floor as her. Her mother still thinks she is curing cancer, which is fine for the time being. She is focused on achieving the goals set out in her proposal and more importantly, getting published. Somewhere in the next couple of years it is likely that her relationship with her advisor will break down. They will stop speaking or even become hostile towards each other. This is especially the case when the student reaches the point where her work does not seem to be progressing well and she is looking for guidance. The guidance she is seeking is not so much advice, but more handholding. This is an extremely difficult period, where her advisor is looking for her to push through her difficulties and resolve problems on her own. Eventually, the two of them will overcome their personal difficulties. Not surprisingly, the relationship will improve as she works through her project hurdles and gets closer to the time of writing her first paper.

As her results mount and the two of them try to assess the appropriate journal for the work to get submitted to, she will also start to think more about 'getting out.' If she is in her fourth or fifth year, her colleagues, friends and family will start to ask her when she is going to graduate. Her answer will change over and over as she works her way through writing and submitting her first paper and thinking about asking her advisor if she can graduate. The green light for this will have to wait until her next committee meeting, but will consume her for months or years before that happens.

Publish or perish

Once her paper is written, polished and submitted to a journal, it enters what is, to her, a black hole. When the paper arrives at a journal, an editor, who is either a full-time professional editor, as is the case for most top biology journals, or an editor who is also a researcher, usually at a lower tier journal, will assess the novelty of the work and the quality of the data presented in it and decide whether or not it is going to be sent out to peer review. The great majority of papers are rejected by top journals without being sent for review. Peer review is the process whereby scientists not involved with the work anonymously evaluate its quality and significance. Up to 80% of the papers never make it past the editor. The leading journals typically reject close to 90% of all submitted papers.

Professional editors are former scientists with PhDs who have left the research lab to take on a career in publishing. These editors actually play an important part in science, even though they will, most of the time, never actually do another experiment for the rest of their careers. The editor must be able to assess science from a wide range of fields and determine whether the work presented in the paper is a significant advance or not. This is easier said than done, and an editor relies heavily upon the peer review process to make the more difficult decisions.

A research paper that is sent out to review is typically seen by three referees, who provide both confidential comments to the editors and comments for the authors of the paper. It takes several hours to write a full review of a paper. The goal of the review is for referees to provide advice on whether the scientist has presented enough data to support the conclusions provided at the end of the paper and to let the editors know whether they think the work is novel enough to publish in a prestigious journal or whether it was an incremental advance; simple enough. But here's the kicker: the reviewers do it for free. In addition to running their lab, writing grant applications and research papers, teaching, counseling their lab members, attending faculty meetings, and traveling to science conferences, researchers give their time to the peer review process. It is widely understood that reviewing research papers and grants is an integral and important part of the scientific process.

Like everything in science, not all is as it would appear. The peer review system can be corrupted by researchers holding a grudge against an author, who are friends with an author, or who are competing with the author. This is why there are multiple reviewers, but in the end the voice of the pickiest reviewer is often heard over the protests of the authors. It's not a perfect system, but no one has really come up with a better one that promotes competition and protects the authors.

Research papers are rarely accepted for publication after a single review. More often than not, the scientists are asked to do further experiments and significantly revise their text before publication. After the reviews are returned to the editor, he or she will confer with other editorial colleagues and decide whether the paper is going to be rejected or accepted, or whether the researcher will be given the opportunity to revise the manuscript and resubmit it. But what is to keep reviewers from making claims or asking for unreasonable amounts of work?

Analogy time: let's say that you are a scientist reviewing a paper and realize that the results are very similar to results that your lab has produced, but which you have not yet published. Let's say that the authors of the paper have figured out a detail that you have not. Do you have your student or post-doc do the key experiments with the knowledge you have gained so you can avoid being scooped? Do you ask for experiments that would delay publication until you can catch up? Do you ask for them to do an experiment that would actually require them to collaborate with you? Do you make such harsh comments that the paper might get rejected? After all, delaying their publication could seriously help you. The reality is that scientists in this position could excuse themselves when they realize that there is a conflict of interest, but many do not. Remember: these are anonymous reviews, so you can say what you like as long as it appears legit. The idea of having three reviewers is to have a balanced view of the paper, so no single reviewer can be unreasonable. The only thing keeping reviewers honest is their own integrity (and sometimes a diligent editor), but it really is a lot to ask of someone who is by nature highly competitive. There is little that anyone could do in any system to completely prevent unscrupulous behavior.

For many scientists, receiving a rejection letter is an enraging experience. This is particularly true when the reason given is that the work does not represent a significant conceptual advance or that it is not novel enough. Talk about a big ol' slice of humble pie. An editor or reviewer doesn't think your work is interesting. After working your ass off for several years, your work is deemed insignificant. The reaction to this can range from rage to depression, but is rarely quiet acceptance. Often a scientist will appeal the decision. However, a paper is rarely published after it has been rejected, and the appeal letter often only serves as a means to vent frustration or anger rather than actually achieving anything.

The pressure to publish can be so great that it leads a young scientist to do the unspeakable: commit fraud. Like speaking Voldemort's name, scientists cringe at the mere mention of someone forging results. It is widely considered the worst thing a scientist can do. A survey came out in 2003 that indicated that the majority of scientists knew of a colleague who had done something that they would consider unscrupulous. Yet the incidence of reporting scientific fraud is very low. The reason? It is extremely difficult to prove that someone has committed fraud unless they do something very stupid or confess when confronted. However, it is so horrible to imply that a scientist has committed fraud that it is exceedingly rare to have someone be confronted and then confess. Not only that, there is virtually no oversight by the NIH to try to detect it. There are of course cases where people do bad Photoshop jobs to try to forge a result, but mostly the data is just manufactured wholesale and this type of fraud is next to impossible to detect. This was the case in 2006, when a Korean scientist was found to have forged data on the cloning of the first dog and the results of deriving human embryonic stem cells from patients with specific diseases. Since science works by piling results on top of one another. Researchers who forge data are destined to be discovered – unless, for some reason, they happen to be correct in the story they made up. But by the nature of science research, data that cannot be reproduced could just mean that the experiment was done wrongly or differently, which makes it even harder to detect fraud.

So, why do some scientists forge data? The answer is unfortunately predictable: the pressure to succeed is too high. For the same reason that some premedical students cheat in undergraduate classes, some scientists will resort to making up their own success to meet the demands of an advisor who is putting pressure on them to get results that fit into their model or a committee that is putting pressure on them to get results in order to graduate, or to avoid the fate of having their funding run out with no papers to show for their efforts. The sense of helplessness that all scientists eventually face is almost never addressed. Forging data is seen as the domain of the unscrupulous and weak.

Let's say that our graduate student has to go back and do some more experiments. She resubmits her paper and it is accepted for publication: the ideal situation. She gets the green light to write her thesis. At the same time, she is deciding if and where to do a post-doctoral fellowship. Once you have your PhD, you are considered inexperienced and unqualified to run your own lab. A post-doctoral position is held up by most scientists as your only reasonable option. It is only recently that students have been exposed to alternative careers after graduate school. The fact is that only around 15% of all PhDs in the biological sciences who graduated between 1993 and 2003 hold tenure track positions four to six years after graduating. This statistic is never revealed to people applying for graduate school or incoming graduate students; not because faculty are unaware of the fact that most don't get faculty positions, but that part of the frame of mind required to be an academic scientist is denial of the notion that you might not get a job. Most of her colleagues will not make it to the coveted faculty position. Instead, they will remain post-docs indefinitely, become teachers, or take a job in the biotech industry.

For a long time (until as recently as the early 1990s) it was considered selling out if a scientist took a job in the biotech industry. The change corresponded directly with the glut of post-docs who just couldn't find faculty jobs and the growth of the biotech industry. However, in the eyes of university professors it meant that you had failed – you weren't good enough to make it. This seems ludicrous today. Now there is sometimes as much competition for a good industry position as there are for faculty positions. Increasingly, there is a large number of scientists going into 'alternative careers,' becoming editors, intellectual property attorneys, policy makers, and biotech business consultants. Indeed, it is now common to have people giving seminars on other career options in the same institutions where professors consider their graduate students to be failures if they don't become a post-doc. The dedication that it takes to become faculty at a research institute is so strong that their students and post-docs not getting what they aim for really doesn't occur to many professors. If there is not a significant increase in the available faculty positions, then fewer and fewer people will go into science as the dirty little secret about job availability becomes more widely known amongst undergraduates.

Our student is one of these people who does not consider the possibility of failure. Graduation and the thesis defense go smoothly, and after revising the text she can officially be called 'doctor'. Now she cleans up some loose ends and is off to become a post-doc.

Becoming a scientist is a more difficult process than most would have imagined. So what! Wait until you hear what a post-doc goes through. Most correctly estimate that getting a PhD in science is hard – you are earning a doctorate after all. (In many other fields, education for example, they hand them out like candy.) But grad school is child's play compared to doing a post-doc. This is where despair and depression are layered like a birthday cake topped with a generous icing of pressure and doubt. Yummy!

Post-docs are usually required to get funding to support their projects. If they don't have any first author papers from graduate school, then they are pretty well screwed. Also, if the new lab they are working in has not yet had publications in the area they are proposing to do work in, they will have a hard time. Many lab heads will require that their post-doc comes in with independent funding before they start in the lab. This means that students start writing grant applications as soon as they are done with their PhD, or even while they are still writing their papers and thesis required for graduation.

It is widely assumed from early in the education of scientists that they don't have to be proficient, clear writers to succeed. The truth could not be any different. While there certainly are a lot of scientists who are as awkward with the English language as they are with the opposite sex, one of the most important tools that scientists can have is the ability to communicate what they have achieved and how they achieved it. The language of science around the world is English, and foreign researchers are well aware of it. What is remarkable is the fact that few graduate programs ever teach their students how to write a grant application or research paper.

Let's assume that our budding scientist has been able to navigate this process and get a grant. (We have thus far assumed that she is really good, so why stop now?) The fact is that the NIH awards post-doc grants to approximately 30% of applicants. Other agencies are a little better or worse, but 30% is normal, necessitating application for several grants to secure funding. Our post-doc will now be getting about $28,000 a year to do science. This is the standard salary paid by an NIH grant to a first-year post-doc. On average, she will spend about five years as a post-doc and typically switch labs and institutions once. That means that she will have spent approximately 11 years becoming a scientist before she is applying for jobs at universities to run her own lab.

After the fire

Amongst scientists there is an often spoken, but little addressed, problem with post-doc life – call it the 'post-doc's lament'. They are overworked, underpaid and often unappreciated. Graduate students produce papers for a lab, but post-docs are the real powerhouse of science. They are expected to perform, but little is done to protect them. Graduate students at least have a thesis committee to protect them from an unreasonable advisor, but post-docs very rarely have a mechanism to protect themselves from their bosses. Policies regarding the treatment of post-docs are developed *ad hoc* and are institute- or even lab-specific. Nightmarish stories are told of post-docs who are pitted against each other in the same lab on the same project. This is done to drive the project, but it results in strife and frustration. Abusive treatment is common and there is little they can do about it. Even at institutions where there are offices of post-doc affairs, little can really be done if a PI is a jerk.

With these issues recognized, the National Academy of Sciences put out recommendations to change things in 2000. There is no evidence that there has been widespread adoption of their recommendations though. In 2002, the national post-doc association was established, but again there is little evidence that it has been effective in creating any widespread change in the current state of affairs. The weird thing is that less prestigious institutions seem to be more progressive in reforming the rights of post-doctoral fellows. There is a great need for transparent, consistent policies regarding the treatment of post-doctoral scientists, but this will not happen unless the NIH and NSF require it of institutions.

Our post-doc doesn't have any problems in this area though. She chose her advisor wisely and will be treated with dignity and respect. After entering her new lab, getting funding and starting her project, she publishes a few really high-profile papers; one in *Nature* and another in *Science*. She has married a post-doc who works in the lab two floors below. He has been there for five years and has published two papers, but they have appeared in lesser journals and he is well aware of the fact that he will not be getting a job any time soon. His initial funding ran out two years ago and despite the fact that he is very good at his job, his project did not turn out to be as exciting as he would have liked. He was therefore forced to move on to another lab and try something new.

She, on the other hand, got lucky with her project. Her advisor has been very supportive, but she only has one year of funding left. Rather than try to secure more funding, she scans the back pages of journals for jobs to apply for. After picking a short list of 47 jobs to apply for, she writes up a description of what direction her work is going in, prepares a CV, writes her 'statement on teaching,' collects letters of recommendation and sends out her application. She finds writing her statement on teaching to be somewhat amusing, because she has not taught a class since her first year in graduate school and generally is speaking from ideology rather than raw experience. She is virtually untrained as a teacher. Each of the jobs she applies to has 300–500 other people applying for them, but because our scientist is so good she gets four interviews. At her interviews she meets lots of faculty, gives a pretty good talk about her work and heads home to wait. She continues to do research and has not heard from the institutions she interviewed at for two months.

Finally word comes and she gets two job offers. This is exceptional. It is more common to get a single offer and be forced to take it no matter where it is or what it entails. The option of waiting a year to apply again, in the hope that she will have another publication to show that her work is still progressing, is more depressing than taking a job that is less than ideal. One offer is in a small department at state school in the mid-west. She didn't like the facilities there and there is a heavy teaching load, but it is a tenure track position. The other is at a more prestigious university on the east coast, in a large department with lots of people that she can collaborate with, but this position isn't tenure track and pays a lot less. The average salary for a starting primary investigator depends largely on the institution and the standard of living in the area. A salary of $50–80,000 is normal, with women making less than men in wide-cast surveys, but our researcher is just thankful that she has offers. Most of her colleagues do not. It annoys her that the other two places she interviewed at never let her know she didn't get the job, but she is well aware of the fact that universities usually don't bother to do this. Callousness is standard. After discussing it with her husband, they decide that she should take the tenure track position despite the fact that the other job sounded more exciting.

Some universities rarely offer tenure track positions. They just can't afford to have any more dead wood than is already floating around the department. This would seem to be a reasonable solution to the problem – get rid of those who are not productive and hire new faculty who will bring in grant money. The result is an active department, but it adds another level of insecurity to the lives of their scientists. They are either unaware of, or just don't consider, an alternate plan. Cold Spring Harbor Laboratory, where I did my PhD work, has an unusual system where instead of giving tenure to faculty, they award something called 'rolling five.' Briefly, when a faculty member becomes rolling five, it means that if they go through a period of low productivity (meaning that they have not brought in enough grant money to keep them in the black), they are given five years to find a position at another university or institute. This seems like a nice compromise compared with getting rid of tenure completely.

In talking with her husband, they decide that they would like to have children in the next five years. She is 35 now and just doesn't want to risk a non-tenure position. At the same time, there is a better chance of her husband finding a faculty position at the same university rather than at the highly competitive non-tenure job. Co-appointments are hard to get. Many post-doc couples try to get them, but more often than not they just find it harder to find a job when a university is faced with hiring both or neither.

There is an interesting phenomenon amongst scientific faculty today that has never really been discussed. Their parents, if alive, are usually still married, and if they are divorced have either remarried or divorced once their children were already adults. If you think about it, this actually makes sense. It takes a lot of fortitude to go through the process of becoming a scientist. One needs to have a pretty strong constitution, and perhaps more importantly you have to have a pretty strong support system to fall back on. Could the results reflect the fact that more people with a higher education have parents that are still married? Perhaps – but it is interesting nonetheless that the situation exists. It certainly isn't an indication that scientists are more grounded than your average Joe.

After accepting the job as an assistant professor – the order of advancement usually goes something like assistant professor to associate professor to tenured full professor – the faculty at her new institute offers her husband a position as well, but not a tenure track position. Instead they offer him a very small lab space as a research fellow (not a professor), but he will have to secure his own funding before he can start. Typically, a university will give new faculty a breather by giving them some start-up money. This is designed to pay their salary and float the lab while it is being established. The money is usually designed to last a few years and is absolutely required to buy lab equipment, computers and chemicals and to pay salary lines for technicians and other support staff. During that time, our scientist is trying to get as much done as she can so she can apply for research grants. The NIH gives some wiggle room to new investigators and requires only a minimum of new data compared with established faculty and requires them to have fewer publications to get a grant. Luckily, she is only required to write the curriculum for, and teach, one class in undergraduate genetics the first semester she is there and to give a single lecture on her work to the graduate students. Just as when she was a graduate student, her teaching responsibility will have to take a back seat to her other responsibilities. Undergraduate students will again pay the price of overworking a scientist.

Our young assistant professor hires a technician to do some of the more mundane tasks in the new lab. She has been thinking about what she is going to do first science-wise, but has not been able to focus on it for several months. As luck would have it, her former post-doctoral advisor was gracious enough to let her take a lot of the materials she developed in his lab with her. This is not always the case. Often, scientists will decide that the project that their post-doc developed was so interesting (and lucrative grant- and research paper-wise) that they compete against them. A sort of bargaining process takes place as a post-doc prepares to leave the lab to forge a path of their own. A new faculty member will certainly have more on her plate than she can reasonably handle. She will be asked to divide her time between organizing her research projects, attending faculty meetings, teaching, writing grant applications, taking on and training graduate rotation students, hiring technicians, attending science conferences and, if she is lucky, taking on her own post-doc. Most starting investigators don't get to take on a post-doc until their second or third year. And oh yeah – lest we forget about her competition outside of her old advisor – she will need to keep up with the mountain of

new papers that are coming out in her field as well. If someone publishes results similar to the results she is getting in
her lab, then she is said to have been 'scooped'. This means
that she will have to come up with even more data and will
probably have to publish them in a much lower ranked journal. This is widely regarded as an embarrassment and really
hurts your chances of getting a grant funded.

Once our scientist publishes her results in a journal, she is
required to share all of the reagents that she made to get
those results with anyone in the scientific community that
wants them, her competition in particular. This is another
system of checks and balances that allows others to repeat
and confirm her experimental results. At the same time, it
removes some of the advantage she has over her competition. The tradition has existed for over a century, but it works
on the honor system. Journals require it, but I have never
heard of a case in which someone has experienced any
repercussions for not sharing their reagents with another lab.
You are basically considered a scumbag if you don't share
them, but many scientists openly refuse to do so. They claim
that too much is invested in the production of the reagents
to share them, but in the end they do know it is wrong. Journals could play a role in policing the policy by noting on their
website that the authors have refused to share reagents, in
effect marking them with a scarlet letter of embarrassment,
but they have taken no action. Sharing reagents after publication is good for science as it allows more rapid progress,
prevents duplication of efforts and makes sure that published results are real, but it also adds another degree of
insecurity.

Currency exchange

While she is struggling to manufacture her career, let's consider how the hell she is going to get a grant. If research
papers are the currency of science, then grants are the life-
blood. You need the currency to afford the blood. The

system has been set up such that the projects with the greatest chance of success are funded. Sounds great right? Let's consider what low risk means in science.

You are a researcher on an NIH grant committee and you have been given the monumental task of reading 30 research grant applications and reporting back what you think of them. After you do that, you go to the NIH and sit in a room with several other researchers who have done the same thing. You will give scores on these applications based on how likely they are to succeed, the degree of innovation, history of success, the institution's ability to support the project and your feeling regarding the applicant's ability to achieve the goals set out in the grant. In the pile there are a handful of applications that have little or no data in them, but a really interesting idea that, if correct, would change the way the field views a particular biological problem. Another set has a series of well-thought-out experiments which progress logically, have a little preliminary data and the researcher has just published a paper related to the topic of the grant in a prestigious research journal. Finally, there is a third set of applications from researchers who have a logical progression of well-designed experiments, have published recently in prestigious journals, and have preliminary data that suggest every single thing that they propose will work out. In addition, at the last minute, they send in an addendum to their application showing that the first of the three objectives they have proposed has actually been achieved. Which application gets funded? The third set, of course. They are going to succeed. They've already proven it. There is such a high chance of success that almost all risk is eliminated.

What does this do to innovation? Novel ideas? It crushes them in a vise of pressure. Very little risk is ever taken in science funding. Recently, the NIH started an initiative that encourages risky experiments, but even here risk has been minimized. The program is a bureaucratic oxymoron. Only well-established scientists are considered for funding of risky projects. These scientists are already well funded and thus are not as inhibited in the projects they can take on to maintain themselves.

It all boils down very simply: research labs have to be run like small farms. Farms run on a very small profit margin and thus little risk can be taken in planting new crops. It is safe to plant corn, tomatoes and carrots. They can be sold. If the farmer plants some broccoli rabe or arugala across a large percentage of his fields, he is risking not being able to sell the fruits of his labor. Larger wealthier farms can take a larger risk. They have the money to gamble on risky crops. Why would you give them an extra monetary incentive to do what they already have the means to do? In the lab, if you have too many people working on risky projects, the papers might not come out at a pace that will allow it to sustain itself. Novel, innovative projects with any degree of risk can make a scientist's career, but more often will end badly.

This pressure has fostered an environment where a lot of shitty science is published in shitty journals. New journals are cropping up all the time. Does the scientific community need them? *No!* There are so many journals out there that even the poorest designed, sloppiest work can eventually make it to press. Rather than having a system where peer review holds the work up to a high standard for all journals, reviewers and editors for crappy journals just let lame science step over a much lower bar. Scientists generally use these journals as a dumping ground that allows them to have another publication on their CV. But that is not the half of it. The pressure to publish is so great that even when a scientist has a fantastic story, what they often do is slice it up like deli meat and try to figure out what is the least amount of data they can put in and still get a high-quality publication – the so-called 'minimum publishable unit.' Ideally, a scientist would be judged on the merit of the work they do, not the number of publications they have. In biological research, there exists a holy trinity of publications: *Cell, Science* and *Nature*. If you publish your results in these journals, you are virtually guaranteed to get a grant refunded, and the companies that run the journals know it.

Cell Press and the Nature Publishing Group have launched a series of more specialist journals over the last 15 years that, since they have the 'Cell' or 'Nature' name associated with them, are thought to be pretty damn good; and largely they are. They also make a lot of money and serve the community well. They make money in three basic ways – personal subscriptions, institutional subscriptions and advertising. In other words you have to pay to get access to what is printed in the journal and, more importantly, get access online. It is a pretty good system. The journals provide a service to the community by overseeing the peer review process, editing papers and providing news on new findings and science policy around the world. Readers pay for it and rightly so. The information is valuable. The money for these subscriptions generally comes from grant money. In other words, researchers use tax dollars in the form of grants to pay to view research results that were obtained with public funding. Not only that, they pay to publish papers in these journals in the form of fees for color figures and page charges.

Recently, there has been a growing movement in the scientific community for free access to this material. The argument is that the companies running these journals are profiting off of an essential component of the scientific process and that the public doesn't have access to research results that they paid for unless they pay for it again. The companies that run these journals respond by saying that they provide an excellent product that serves the community well and that they should be rewarded for it. They are, after all, operating for profit. They also argue that the profit they make goes into supporting less profitable publishing ventures that also serve the community well. This is only partially true though. They do funnel money into publishing ventures that cost rather than make money, but these ventures are intended to be profitable in the end. Otherwise they wouldn't do it.

Some enterprising scientists have now spearheaded the launch of a series of journals that are free ... sort of. Anyone, even non-scientists, can view the content. In addition, many other journals make their content free after a period of time, generally six months after publication. The only problem is that the business model for these journals will not work if widely applied. The funding model for these free journals falls apart pretty quickly when one considers how to publish very specialized journals that serve a small subset of the research community.

Take for example one of these ventures, the Public Library of Science (PLoS). When this venture was launched they hired some seasoned and respected editors to launch a series of journals, starting with *PloS Biology*. They obtained a $9 million private donation to do so and presented their publishing model to the community. Researchers pay $2500 to have their paper published in the journal (assuming it passes a rigorous peer review). The journal saves money by having its content primarily online and will provide subscriptions to a print edition at cost. Therefore, for their business model to work, they will need to publish a certain number of papers per month to cover their overhead. This shouldn't be a problem, because biology is a very general topic and there is a lot of great research to publish. What if it were a more specialized field, like genetics? We'll soon find out. They have now launched several new titles, including *PLoS Genetics, PLoS Pathogens, PLoS Neglected Tropical Diseases* and *PLoS Computational Biology*. Our young scientist, in a small department, would have to take a large ideologically driven risk with her and her post-doc or grad student's career to publish in these new journals. It is an experiment in publishing that could, like many science experiments, fall apart.

Zealots within the movement are quick to demonize any company that makes money in science publishing, but if pressed would be forced to admit that their own model would not work on a wider scale. In any other business, the idea that profit is bad would be laughable, but scientists are an idealistic bunch. Time will tell, but it is highly unlikely that for-profit science publishing is going to disappear unless a ridiculous amount of money is funneled into the idealistic model – public money that is.

The Bush administration has now initiated a policy that if publicly funded scientific results are to be published, then the NIH 'requests and strongly encourages' scientists to make the final edited text publicly available. To do this they have launched a program where scientists submit their papers to an NIH database that the public can access.

What is the best they can hope for: access to science results that less than one per cent of Americans can understand? I know it sounds elitist to say that the public will not benefit from free access to research they are paying for, but it is the truth. Most Americans don't know the difference between DNA and protein. How the hell are they going to wrap their heads around science papers that are complicated for people with years of training? The idea that this initiative will serve some lofty public good is completely naïve.

The new policy has opened several other cans of worms. First, there is no oversight in place to prevent multiple versions of research papers appearing. While the NIH 'encourages' scientists to submit the final edited version of their text, the process is frequently screwed up. The policy also ignores the inherent value that publishers play in editing text and the fact that it costs them money to get those papers into shape for publishing. Thus there is no real regulation of what scientists can write in the paper that they submit to the public database. It also allows researchers to make claims that their data does not justify and put it in the public domain. These claims are usually removed during the peer review and editing process,

sex, drugs and dna

but who is to stop a researcher saying whatever they want in the version submitted to the NIH database? Personally, I can't wait for scientists to start submitting their papers with statements about how the administration has shown contempt for science and do not understand how science works.

In the case where multiple versions of the paper appear publicly, serious intellectual property issues could arise. Universities and companies will have to shell out lots of money to square these problems. Imagine the arguments over which version is valid. What a total mess.

The policy is actually a watered down version of the original proposal that would have *required* that publicly funded science be submitted to the database, but small publishing companies and scientific societies who publish journals and rely on subscription fees to remain solvent placed a lot of pressure on the NIH, claiming that a mandatory database would put them in serious danger of going out of business. In the end, the policy is a eunuch and a total waste of money. It is just a suggestion, and few scientists actually view it as valuable and thus do not participate in it. In 2006, it was revealed that less than 4% of eligible papers had been submitted to the database, that the majority of the papers in the database were not supposed to be in it, and that authors were frequently submitting papers before the journals' embargo on them had expired. None of these cold realities has stopped the administration from claiming that the database is an actual archive of NIH-funded research, that it will be valuable for scientists trying to search for research papers, that it will allow the NIH to manage its research portfolio, that it will actually provide scientists with higher visibility for their work, that it will improve the public understanding of biomedical research, and that it will strengthen the impact of research findings. None of this is true and the NIH, their scientists and the greater scientific community know it. I cannot for the life of me think of a single previous instance in the entire history of the National Institutes of Health where they actively tried to deceive the public.

While publishing provides the bread and butter for a scientist, there is another component to science that provides a healthy dollop of drive (or at least it used to): the scientific conference. It used to be that scientists would get together with their colleagues from around the world to discuss their latest results, forge collaborative efforts on scientific problems to which each side could contribute results or expertise and get a bird's-eye view of what's going on in their field. The importance of the scientific meeting cannot be underestimated even today. But things have changed a bit. The speed with which science happens is at once a crawl and a race. It takes a long time to put together a story that makes a significant conceptual advance in biology, but if you miss a step, someone will beat you to the punch. Over the past few decades, the scientific meeting has changed dramatically, and it is now more common for scientists to publicly discuss results that are already or are about to be published. Time after time, scientists have been burned by discussing their results publicly because a competitor, who has similar results, is sitting in the audience. Thus the original spirit of the scientific meeting is dead.

Let's say that our young investigator is presenting the work from her new lab at a science meeting. She is eager to show that she is really working independently of her former advisor and shows all of her latest data. Here are three scenarios that she will have to consider. Her competitor is sitting in the audience, fearing that he is about to get scooped. He immediately submits similar results to a journal for publication and hopes it is not too late. If they are friendly, they might get together and try to publish two papers back to back in a journal, but revealing that you have similar results has its downside as well. What if they don't want to cooperate?

Scenario number two often occurs if the competitor has conflicting results. Rather than sparking a lively discussion that advances the field, it will likely result in a race to publish. It is very hard for scientists to admit they are wrong, even when it is pretty well proven. So, out into the public domain go two results that are in complete conflict and the parties involved often know it. Publication usually takes priority over clarity in the field and if conflicting results are already out there, she will have a much steeper hill to climb to get her results published, having first the task of proving that her competitor was indeed wrong.

Still worse is the third scenario, in which the competitor goes back to his lab, does the same experiments and publishes while our scientist is trying to polish the story. There aren't a lot – percentage-wise – of these scumbags, but they are out there and it only takes one to sour an entire field. After thinking it over, she prudently bows out and shows some of the work she published the year before and speaks of her current work in the vaguest of terms.

If you want to hear the latest results from someone's lab, it is best to talk to them in the bar after their presentation is over. In fact, some of the most productive discussions regarding science take place in bars and restaurants in the evenings after the scheduled sessions are over. Most scientists can tell you a story about how they figured out a problem, designed an experiment or forged a fantastic collaboration after tipping a few too many. Ethanol definitely lubes up the noggin. Luckily, our scientist is fond of the drink. Realize that grant money – your taxes – pays for scientists to attend these meetings. There is no available estimate of how much money is spent on airfare, hotel rooms, food and attendance fees for scientific meetings, but when you consider that there are companies that specialize in running them, you can guess that there is some significant money being spent.

With the road to becoming a scientist being paved with so many potholes, one might think that it would be hard to find people to take on these challenges. This is simply not true. Part of the reason is that it is not made clear to students entering graduate school that they will probably not make it to a full professorship. In fact, there have been a number of reports that indicate that we are not producing enough scientists in the US. The problem is that the original report from the National Science Foundation predicting a shortfall of scientists in the US was based on an expected wave of retirement that never happened. The number of unemployed scientists in this country is rising. Even with a doubling of the NIH budget during the Clinton era (something that President Bush has implied he had something to do with on several occasions), the number of new jobs for scientists did not keep up with the number of scientists on the market. Not only that, there is an increase in the number of foreign-born scientists competing for jobs in the US. The number of years that one spends as a post-doc is increasing. Simply put, postdocs are wondering when this 'shortfall' is going to result in a job for them.

With a 2001 report on national security reporting 'current trends of supply and demand ... may seriously threaten our long-term prosperity, national security and quality of life,' there is a suggestion of a crisis that is simply not there. The claim that mismanagement of science and education poses a danger that is 'second only to weapons of mass destruction detonating in an American city,' is merely a scare tactic; what's more, if this is really a threat little is being done about it. A closer look reveals that the predicted shortfall is based on numbers that reflected a drop in the rate at which graduates were being produced, rather than on an estimate of the number of jobs that would go unfilled. With several hundred applicants for a typical faculty position, it would seem that students are being grossly misled about their prospects for employment and that the government is continuing to scream that we don't have enough scientists. It is a nauseating display of misguided intentions, and smart talented people are paying the price.

On her 37th birthday, our scientist realizes that the ticking sound that has been keeping her up at night for the month leading up to her birthday is her biological clock telling her that she really wants kids ... *now*! She and her husband have discussed it over and over again, ending each conversation with the same rational conclusion: that they will start a family when things 'settle down.' Her career has kept her from having kids up until now, and reality has finally set in. She has been all too aware that, as she gets older, the chance of her getting pregnant and having an uneventful pregnancy decreases. She has students, post-docs and technicians who rely on her to run the show every day. There is no good time to have a child. There is no real maternity leave for scientists either. While officially universities will allow professors to take leave, the reality of it is that you are more likely to find an infant in a professor's office than a professor out on full maternity leave. Either way, she will work even if it is from home.

If you are considering having children as a post-doc, you risk falling out of favor with your boss. Some scientists used to flat-out ask women post-doctoral candidates if they were planning on having children in the immediate future. A few lawsuits later they are more savvy than that, instead trying to beat around the bush to the same end. This aspect of science is not unlike the discrimination of women that is seen in the business world.

I have given you a taste of what it takes to become a scientist, and the less than glamorous life a scientist leads. With all of the complications associated with becoming a research scientist: why do they do it? In the end, the reality is that scientists are members of a cultish band of men and women with an unusual thirst for discovery.

Let's now consider some of the most controversial issues that scientists face today: the wide misunderstanding of, and uneducated ideological opposition to, what they do – and the misrepresentation of what they are achieving.

2

stem cells and cloning

"After trying for years, Doris and Stanley turned to reproductive cloning..."

No, we all can't just get along

There are three topics in modern science that have caused more controversy than stem cells – the development of nuclear bombs, the notion that the Sun is at the center of our Solar System and the theory of evolution. Stem cells have little in common with the bomb, but there is a fundamental similarity between the stem cell controversy and those surrounding evolution and the celestial body at the center of the Solar System: the people who are offended by them. If you happen to be a creationist, believe that the world is less than ten thousand years old and that evolutionary theory is a conspiracy put forth by scientists and/or Jews who hate Jesus, then by all means put down this book and take a moment to pray.

It comes as little surprise that those who wear their religion on their sleeve have a problem with science. This has been an unfortunate reality from at least the time of Copernicus and Galileo, whose books on celestial observations were placed on the Roman Catholic Church's List of Forbidden Texts. Galileo was actually imprisoned for endorsing Copernicus' theory that the Earth rotated around the Sun. It only took the Church until 1998 to apologize for their treatment of him. I'm sure his family feels much better now.

More modern examples include opposition to organ transplant, blood transfusions, *in vitro* fertilization, recombinant DNA technology and now stem cells. Unfortunately, those who stand in front of discovery with Bible in hand also happen to be amongst the most vocal and motivated in society. The energy for such hateful rhetoric seems to come from the sincere belief that scientists are challenging Biblical scripture and are destroying human life. They truly believe that embryonic stem cell research is inextricably linked to abortion. In other words, scientists are baby killers. Since we have a handful of vocal religious extremists in leadership positions in the US, the debate about embryonic stem cells and embryonic cloning has been driven to the level of presidential debates, the floor of the US Congress, and the United Nations.

Stem cells and cloning are two very different things. A lot of the confusion comes from the jargon term 'cloning' which has been popularized in movies and television to mean the creation of a carbon copy of an animal or person by a mad scientist with funky hair and a loud and rather evil laugh. Not only are they different issues, the controversy around each is quite different. Stem cells are exactly that – cells. However, these cells are the magic beans of the body: they keep dividing and dividing, in essence constantly renewing themselves. Cloning is a technique that can be used to generate an embryo that is almost genetically identical to a living person. The two issues intersect at a single point, which has been used by conservatives to demonize both issues. Before I completely lose you, lets do a quick 'Stem cells 101' so we're all on the same page ... except the creationists, who are busy talking to Jesus.

Mommy ... What's a stem cell?

Most cells in your body are not actively dividing. They have matured into a specialized type of cell, a liver cell for example, and will do their job and die. Stem cells have an important job – they make more cells. They do this by continually dividing and replacing both those cells in your body that have kicked the bucket and themselves at the same time. When they divide, one of the two cells stays a stem cell while the other matures, does a job and dies. Simple enough. There are many kinds of stem cell, so the term is usually preceded by a descriptor, like blood or liver. Much of the debate has hinged on the relative value of adult stem cells vs. embryonic stem cells for developing therapies to treat disease. The arguments arise from a fundamental lack of understanding of what these cells are capable of doing.

In an early embryo (I'm talking about a 14-day-old cluster of cells) there is a balled-up structure called the inner cell mass that contains embryonic stem cells that have the potential to become any type of cell in your body. They are said to be pluripotent. A common mistake is to say that they can become *any* type of cell or that they are 'totipotent,' which is not true. They cannot be put into reverse and become the other type of cell in that 14-day-old embryo or the cells of an embryo that is 8 days old. They also cannot become a cell type that requires them to have different genetic material than they already have; for example, human embryonic stem cells cannot become a dog's testicle cell. The point is that they come from human embryos and can become most human cells.

Stem cells that come from later stages of development are appropriately called adult stem cells. These cells can become the type of cell in the organ from which they were harvested – blood cells in the case of bone marrow stem cells. There are a few reports that say that some adult stem cells can be coaxed to turn into cells from organs that they did not come from. This is scientifically controversial, but is precisely the goal of a lot of adult stem cell research. If you can turn blood stem cells into heart muscle cells, then you might be able to inject them to repair a dying heart. Adult stem cells have actually been used in therapies for Severe Combined Immunodeficiency (SCID) and a number of other human blood diseases. For the treatment of SCID, for example, bone marrow stem cells were harvested from a SCID patient, were corrected for a genetic defect and reintroduced into the marrow. This is not a widely applicable therapy for human diseases though, because generally, as the body develops, the stem cells themselves become more specialized and thus their ability to differentiate into different types of cell becomes more limited.

Embryonic stem cells, on the other hand, are obtained from the inner cell mass of a very early stage embryo. Once extracted, they are grown in dishes and so-called cell lines are created. A stem cell line is a group of cells with the same properties that can grow in a dish and divide endlessly without turning into other types of cells unless they are directed to. A good analogy for thinking about cell lines is to think of them like cars – a 1975 Mustang vs. a 2006 Mustang for example. The 1975 Mustang is made in the same factory, by the same company, and the same people as the 2006 Mustang. However, the 2006 model is made on a more efficient assembly line with better components and a more economical engine. So, while they are both Mustangs, they have very different properties.

Embryonic stem cell lines are derived from a single cell from an early embryo or a cluster of cells and are useful for scientists interested in studying how cells turn into adult cells during development. They can also be used to develop medical therapies. The key here is that the embryo is destroyed in order to extract the cells and establish the cell lines necessary for research. There are no two ways about it: currently, when cells of the inner mass are extracted, the embryo is destroyed. There is ongoing research into ways to extract cells from embryos so that they are not destroyed, which is appealing to many Americans. But as you'll see, the logic of even developing such a system is questionable.

Human embryonic stem cells were first isolated in 1998 by researchers at the University of Wisconsin, and almost immediately anti-abortion groups denounced the work. The controversy grew from that point on, with religious groups denouncing the work as a fundamental danger to American ideals and the sanctity of life. Researchers remained relatively quiet on the subject because none of the arguments being made against their work made much sense. Basically, they figured that the public would see right through the unsophisticated arguments being made at that time. Slowly it got worse, and scientists started refuting the claims made by these groups. This set up the public debate that the religious right was hoping for. Scientists are not equipped for or generally inclined to form aggressive rhetorical grass roots campaigns. So a lot of the pieces showed up in science journals, while the right turned to divisive media-grabbing pronouncements.

The public debate came to an ugly head on 9 August 2001, when President Bush came on television to address the public on the issue. People on both sides of the issue sat glued to the television waiting for the President's verdict. In his address he pronounced that he would restrict federal funding for embryonic stem cells to 'more than 60 genetically diverse stem cell lines,' that were currently available. Further, he stated that federal funding could not be used to create more embryonic stem cell lines. He specifically stated:

> Leading scientists tell me research on these 60 lines has great promise that could lead to breakthrough therapies and cures. This allows us to explore the promise and potential of stem cell research without crossing a fundamental moral line, by providing taxpayer funding that would sanction or encourage further destruction of human embryos that have at least the potential for life.

The policy was an unexpected compromise right down the middle. Conservative groups were unhappy because the President did not ban embryonic stem cell research completely, and scientists were both relieved and annoyed. They were relieved that he did not ban the research completely as the Christian right had hoped for, but annoyed that research policy was being dictated by a conservative agenda. What most don't realize is that scientists had no idea how many lines were currently available for research. Even those who work on stem cells were surprised to hear that there were 60 lines available. The reason is that there was no central repository for the cells at the time and no one had ever really tried to count the number of lines. The reason? Sixty cell lines is not very many and they were held in freezers around the world, many in private hands. So there was no need to count or for a centralized collection. Also, the scientists were well aware of the fact that the field was new and that the available lines were not very useful for most purposes. So counting them would have been seen as a silly activity at this stage of the research. It would also have been seen as preposterous for any researcher to claim that their cell line would be useful for developing a cure for disease. Yet, here was the President of the United States making exactly those claims.

Scientists immediately asked where the President got his magic number and who these 'leading scientists' were that said that these lines were enough to do research that could lead to therapies. As it turns out, Mr Bush underestimated the actual number of lines that were eligible under the new policy. The actual number was 71. But there is a huge difference between eligible and available. There weren't 71 lines available or 60 cell lines available: the number was more like 2 when he made his speech. Had administration officials called a few stem cell researchers, they would have known that. After the President's pronouncement, the 'real' leading scientists who actually work with embryonic stem cells were forced to explain why 60 cell lines are actually a small fraction of the number that would be required to come up with actual therapies. So now the President has forced researchers to explain the intricacies of biological research to a community that is ill prepared to understand what chromosomes and cells are, much less the nuances of research. To their credit, they did and were quite successful at it – and that is where the liars came in.

Everyone from the conservative talking head pundits to Republican congressmen and religious leaders just made up facts about what is required to do stem cell research, about how 'experts' told them that it was doomed to fail, how this was killing babies, and how adult stem cell research was far more likely to be effective in curing disease. You name it, conservatives said it. They did absolutely everything they could to discredit research without having an understanding of it. It was shameful and sad.

. After Bush screwed up he did what any ineffective politician would do when dealing with something he doesn't understand: he threw money at it. Today, according to the NIH stem cell registry, there are 22 stem cell lines available for research using federal funding. Tens of millions of dollars are being spent on research into these cell lines, the majority of that money being used to show how completely useless they are. Researchers are actually taking their federal grant dollars and using it to show how bad the stem cell lines that the President said would cure people of disease are. Not only that, few of the lines were derived from a single cell, meaning that the majority of them were derived from a mixed population of cells. This is less than ideal for therapy. In addition, all of the cells need to be fed in culture dishes by a thin layer of mouse cells. When human cells are fed this way they can become infected with mouse viruses, making them very dangerous for therapy. And that is precisely what happened: every single one of the stem cell lines that the President authorized funding for is infected with mouse virus, making them completely useless for human therapy.

In September 2005, researchers at Johns Hopkins Medical School showed that the President's stem cell lines had actually acquired DNA mutations since they were originally created. This type of mutation accumulation happens normally in many different types of cell line. This is because every time a cell divides, opportunities for new mutations occur. So as a cell line gets older, the odds of it acquiring mutations in genes increases. It was previously thought that some stem cells might be resistant to mutations, and that still might be the case, depending upon where, when and how they are made. But to figure that out, researchers need to look at a lot of stem cell lines. Some of the mutations found in the President's stem cell lines have actually been associated with cancer previously. Using a stem cell line with these mutations in them to develop a therapy is absolutely insane. So the President has scorned the scientific community and wasted tens of millions of taxpayer dollars on a handful of stem cell lines that are absolutely useless for anything but showing that the President should not be making decisions on how science is done.

Looking back at the President's decision is truly telling. Besides lying to the public, the truly strange thing was that the President was making science policy through a speech. Since his address, there has been no formal White House policy written on the matter. It was the speech and the speech alone, with radically inaccurate details, that created the policy. This was indeed the first time that any President had limited the ability of biologists to explore new therapies in a speech.

Some will say that my take on the available stem cell lines is too critical, that we can learn a lot about stem cells from them. Fair enough. Let's take a look at the details. The NIH currently charges $5000 for a single vial of these cells. These early cell lines are actually very hard to grow, and if you screw up, you can't get another vial unless you pony up another $5000. You could take one of the classes available to teach you how to grow them, but severe restrictions have been placed on biologists working with the cells that limit what they can do with them. For example, scientists who buy the cells can't share them with collaborators. This rule has to do with intellectual property limits on their use placed on them by those who own the lines. Most people think that the government owns these cell lines, but it doesn't: it just distributes them. This is significant, since many labs do not have the full array of expertise required to achieve their research goals. Collaboration has become the norm in science, and this one limitation completely upsets the normal methods that scientists use to address problems. This is particularly unfair to smaller research labs, which are often run by young scientists. If a scientist treats the cells in a certain way or alters them genetically, they technically can't send them to other researchers to evaluate. Collaboration is common in fields where the government has not restricted research, but not in stem cell research since the President got involved. This leaves us with federally funded cells that are arguably useless, difficult to work with, very expensive and have use restrictions placed on them. Let's get back to the debate.

Conservatives have been trying to make the point that adult stem cell research was far more promising and far more advanced than embryonic stem cell research, and thus scientists should drop this quest to work on embryonic stem cells and work solely on the adult cells. This is partially true. Adult stem cells were isolated more than 50 years ago and have indeed been better characterized than human embryonic cells, which were only isolated in 1998. Duh! They are currently being used in clinical trials and appear to be working for some diseases. But, from the start, we know that it is unlikely that these cells will be useful for brain diseases, kidney disease, metabolic diseases like diabetes and a host of other diseases because of the inherent limited potential of the cells. Remember: the older the cell, the more limited it is.

This brings up a question that is really not discussed – Why did it take so long for scientists to isolate embryonic stem cells? After all, if they have so much potential, wouldn't the scientists of the world have found them earlier? The truth is, you need human embryos to isolate embryonic stem cells, and human embryo research has always been a touchy matter. Where do you get the embryos? Well, the best source for these embryos is frozen spare *in vitro* fertilized embryos – those that are not going to be implanted in a woman to make a baby. *In vitro* fertilization only became a viable way for infertile couples to have a child in 1980, and the federal government has never funded research on these spare embryos. They are to remain frozen until such time that the parents or owners of them say that they do not want to keep them any more, at which point they are destroyed. The destruction of these embryos is not well regulated and can happen in a number of ways; just leave them out at room temperature until they die, pour bleach or some other chemical in the tube to destroy it, or put them in with medical waste that is run through a machine called an autoclave that exposes them to superheated steam and pressure. The machine is commonly used to sterilize equipment. In other words, it is legal to flick a tube of these frozen embryos into the trash can, but not legal for NIH-funded researchers to use them for any kind of science.

It is important to understand the scale of hypocrisy as well. In 2003, a RAND study came out that estimated the number of frozen embryos in the US to be around 400,000. While there are no completely reliable studies on the number of frozen *in vitro* fertilized embryos, the RAND study is considered the best so far. What is not pointed out by most people is that many spare embryos created during *in vitro* fertilization are never frozen; if they are not implanted, they are discarded immediately. So there are potentially a lot more embryos out there that could be used for research. This has not stopped congressmen and religious leaders from claiming that scientists are planning on subjugating women by paying them for their eggs.

Another fine point that the public is unaware of is that the live birth rate per frozen IVF embryo transfer is only 24%. This is probably because embryos do not survive the thaw or because, despite the fact that they are frozen, they still degrade over time. This phenomenon is well known to scientists who work with frozen animal embryos, but seems to escape the public debate. If conservatives really think that embryos should be considered human lives, then they should also be against the entire IVF practice, especially the practice of freezing embryos, because it is not medically necessary and results in some embryos dying during the process. The reason for freezing embryos is that it prevents women from having to go through multiple invasive procedures to harvest their eggs in the event that they do not become pregnant after the first round of implantation.

There are exceedingly few people who place maximum value on a fertilized embryo, and we all know it. Just apply the old philosophical trick: start a fire. You wake up one morning and you realize that you are a creationist and that stem cell research and use of human embryos in research is wrong. Refreshed from your epiphany, you decide to go for a walk and enjoy the world that God and baby Jesus provided for you. A few blocks in, you come to a burning building. A closer look reveals that it is a fertility clinic and that there are people inside. Mustering all your courage, you run inside to save them. Once inside, you see a woman coughing, over-come by the smoke, lying next to a container that says 'frozen human embryos' on it. What do you do? In one swoop you could potentially save hundreds or even thou-sands of embryos, or you could save the woman. Who would really save the embryos? Let's say you decide to save the woman, but still believe that embryos are human beings, but with slightly less value than people who have actually been born. Do you run back in? Don't feel bad for saying no – although embryos have the potential to become human beings they really shouldn't be afforded all the rights of humans. They don't have thoughts, feelings or ideas because they don't have brains, neurons or any real differen-tiated cells for that matter.

Pain sensation does not develop until the 28th week of pregnancy: that's seven months in! Yet many people buy the rhetoric that embryos feel pain upon abortion. It's simply not true. This did not stop Senator Sam Brownback from intro-ducing a bill that would require physicians to discuss fetal pain with women seeking an abortion and offer them anes-thesia for the fetus. But let's be clear: the IVF embryos we are talking about have never been inside a woman and are only 14 days old. They were conceived and remain in dishes and tubes.

Most people don't realize that President Bush's shiny new policy did not change any of the funding rules for isolating new embryonic stem cell lines. Scientists have never been allowed to use federal grants to harvest stem cells from human embryos. The original policy of not funding research on *in vitro* fertilized embryos actually served to delay the first isolation of human embryonic stem cells for close to two decades. The policy is hypocritical at best and foolish at worst. Private funding by the biotech company Geron Corporation was used to isolate the first cells, and indeed private funds are still being used to fund the creation of stem cell lines. However, these new lines, once established, cannot be used for research funded by the NIH or any other government agency. But that has not stopped the President or conservative activists from claiming that they don't want to fund research that results in embryos being destroyed (even though they will be destroyed anyway).

Even with the realization that he goofed on the number of lines that were available, President Bush did not change the policy so that new lines created with private funds can be used by federally funded researchers. Instead, the government has gone ahead and earmarked funding for grants and 'centers of excellence' that will focus on the useless 22 cell lines created before the 9 August speech, even though these lines were also created with private funds.

Since the speech, hundreds of new embryonic stem cell lines have been created at universities across the world. Scientists are slowly coming to understand the best ways to isolate the cells and culture them. It seems only logical that they would get better at it, and the quality of the lines would increase as they did. So, lines isolated in 2007 are indeed better than those isolated six years earlier with cruder techniques. They no longer depend upon mouse cells to feed them, they are easier to handle and culture, and they are far better suited for the development of therapies. So the science is advancing, but not at the rate it could because of hypocritical policies. It's like forcing us to light a fire by rubbing two sticks together rather than using a match or a Zippo. Put

> I'm sure there's a lot of people frightened ... biotechnology is a long word and it sounds ... they may say, well, I don't know if I'm smart enough to be in biotechnology, or it sounds too sophisticated to be in biotechnology.
>
> *President George W. Bush, 7 November 2003*

another way, our scientists are not using the latest and best resources to try to cure disease. If you asked the average person on the street how they would feel if their doctor was revealed to be not keeping up with the latest drugs and therapies, they would find another doctor. And that is what will happen to scientists who are looking solely into the Bush-funded 22 stem cell lines: their work will be as obsolete as their starting material – the doo-doo in, doo-doo out principle.

Why are scientists doing this work anyway? Why not just move on? Surely there are other ways to skin the disease cat? It would certainly be easier not to put up with rhetoric, vilification and red tape. What stem cell researchers are going through makes normal academic tribulations look like an insouciant dreamscape. There is a simple principle in science: discovery can only be directed so much. Scientists roll out X-rays and penicillin as examples of discoveries that were made fortuitously, but the reality is that fortuitous discovery is the norm in science. Take a look at the Nobel prizes awarded in medicine. More often than not, Nobel laureates could not have fathomed the impact of their research while they were doing it. Scientists follow leads in the hope of finding something that fits together the pieces of a puzzle, but are rarely directed towards greatness out of pure insight.

Thus, to reach greatness in science, one often has to do things that on the surface do not seem directed towards lofty goals. Having politicians tell scientists the best way to do their job is really offensive, as offensive as telling generals the best way to plan a battle. So the answer is that scientists do it because their interest and intuition tell them to.

Attack of the cloners

When the public are polled about their opinion on the performance of different government institutions, they answer that they favor the job being done by the CDC and NIH over just about every other agency. In modern history, there is only one movement in biology that has been ultimately regrettable: eugenics – the idea that selecting for desired traits in a population is a good way to avoid disease in the future. This led to sterilization of the mentally disabled and a host of other atrocities. In practice it was a really bad idea. Other than that, the checks and balances in basic science research prevent horrible things from happening. Clinical research and medicine are another story, but basic biology research is pretty golden. So there is reason to trust biologists, but that argument holds little weight in a society that thrives on mistrust and paranoia.

I mentioned at the beginning of this chapter that embryonic stem cells and cloning are two different things linked by a common group of detractors. That was an oversimplification, just to get you thinking of them as separate entities. The thing is, cloning is really just a generic term used to describe a number of techniques that researchers use. For example, DNA cloning is a method to make many copies of a piece of DNA. It is performed by inserting a piece of DNA into a circular DNA 'vector' that is then stuck inside bacteria, which then divide and create more of the desired DNA. The human genome could not have been sequenced without DNA cloning. For that matter, almost every modern biology lab in the world uses some form of cloning.

The controversy surrounding cloning is somewhat simpler than the stem cell issue and actually rotates around a single technique called somatic nuclear transfer. Before your eyes roll to the back of your head, this is not a complicated idea. To do it, you take an egg, suck out the nucleus and replace it with the nucleus of another cell from the body. What you have done then, in essence, is create an embryo.

The normal way in which an embryo is made is that sperm and egg fuse together. This creates an embryo that has twice the DNA of the egg or sperm alone. The process is called fertilization. This is precisely where the confusion lies. Fundamentalist Christians consider this the point of conception, whether it takes place in the womb or in a dish. However, the medical community considers the point at which the embryo implants into the womb to be the point of conception. That is because more than half of eggs that are fertilized never implant in the womb, but are flushed out when a woman menstruates. The fusion of sperm and egg activates the single-celled embryo to start dividing.

In somatic cell nuclear transfer, you bypass the need for sperm and egg to fuse, and instead of having a combination of DNA from mother and father the new embryo has a single contribution from one of them. This embryo then starts dividing and is considered a clone of the individual that the nucleus came from. This is the very same process used to clone Dolly the sheep in 1997. In that case the nucleus that was injected was from a mammary cell; thus the lamb was named after country singer Dolly Parton, who is known for promoting herself with more than just a catchy melody.

There are two basic controversies in cloning: first the creation of an embryo by otherwise 'unnatural' means, and second what you do with the embryo once you create it. With what I have told you about the technique of somatic nuclear transfer in mind, we can deal with both of these issues.

Creating embryos through somatic nuclear transfer is a necessary step towards implementing a technique called therapeutic cloning. The idea is that if you are going to use embryonic stem cells for therapy, then you want to make sure that they are not going to be rejected from the body. One way around having to use anti-rejection drugs is for the cells to be genetically identical to those of the patient. So you make a cloned embryo from that patient, harvest the stem cells from it, and coax them to develop into the type of cell you need to cure the patient: pancreas cells for those suffering from diabetes, or brain cells for Parkinson's or Alzheimer's patients. If the person's disease is due to a specific genetic defect, then the defect could be corrected in the cloned cells and then implanted to cure the disease.

Finally, if one is going to study the development of a disease in human cells, it would be a significant advance to be able to do so in stem cells from a patient with the disease. Therapeutic cloning is just an idea at this point, a great idea. The technical details still need to be worked out.

The question is intent. There is at least the potential for an embryo created by somatic nuclear transplant, if implanted into the womb of a woman, to develop into a human being, in essence creating a cloned human being *à la* Hollywood movies. Some argue that the single-celled embryo is in fact a human being, but for our purposes, I am going to avoid the conservative rhetoric about what to call embryos. For simplicity, 'human being' will refer to a baby after it is born. The key is that the embryo has to be implanted into the womb of a woman for this to happen, a practice that is not illegal in the United States. This is not some trivial technical fact.

To clone a human being, you first have to find a doctor who has both the expertise to make a cloned embryo and a willingness to implant it into a woman. Second, you have to find a woman who would be willing to have a cloned human embryo implanted in her, a thought that rightfully scares the hell out of just about every woman. Third, the embryo has to develop normally. Many scientists have argued vehemently that these cloned embryos could never develop into a human being because of the complications that have been observed in cloning other animals. Some have gone so far as to say that making a cloned human being is impossible.

Indeed, cloning animals is not easy. Creating the initial embryos is not the problem; it is what happens to them during gestation that is frankly horrible. It currently takes several hundred implanted embryos to get one cloned animal. Most of these embryos die quite early, probably because they do not go through the type of DNA reprogramming that normally happens when sperm and egg fuse. This reprogramming is necessary to control which genes are switched on during early development. If the nucleus that is transferred to the egg in cloning does not go through reprogramming, the embryo dies. Those embryos that go through partial reprogramming and make it to adulthood often suffer from 'large offspring syndrome' where some of their organs and placenta are enlarged. Many cloned animals also suffer from metabolic defects and overall have a pretty grim fate. It is not clear if complete genome reprogramming ever happens in cloning. Thus, implantation of a cloned human into a woman is universally thought to be dangerous to the mother and dangerous to the child. This doesn't even begin to get into the moral issues that surround the very idea of stepping into Aldous Huxley's nightmare *Brave New World*.

Ideological opponents of all cloning techniques would have you believe that scientists have collectively lost their heads – that they have not considered the ramifications of their work and that it is their job to remind the world of what ills science can bestow. This is a steaming pile of shit. These people regularly misrepresent science and cherry pick from the scientific literature to suit their agenda, yet since they usually have religious affiliations it is considered bad form to call them what they are: charlatans. They take money from well-intentioned people who have been bilked into believing that the world is falling apart at the seams because of the actions of rogue scientists and the exposure of one of Janet Jackson's tits on live TV. These people have been around the science of stem cells long enough and have been exposed to simple explanations time and time again to the point where you can be sure that they are willfully misrepresenting it.

No wonder scientists are angry at them. The cardinal sin of science – making up data or misrepresenting facts – is enough for scientists roast their own on the spit. Imagine how they must have felt when the charlatans brought their snake oil to town and started misrepresenting legitimate research findings.

Why do scientists push the boundaries of possibility? Simple: through the mandate of most science funding agencies, they are investigating the cause and searching for cures for disease. But more precisely, scientists discover for a living, and conservatives often view that as a bad thing. But instead of just coming out and simply explaining their concerns, they have been programmed to attack what they don't understand. This allows them to start the scare machine and keeps people in the flock with a common goal of defeating an imaginary evil. As for scientists: take a bunch of people with a thirst for discovery and unleash them on a problem and truly amazing things will happen – plain and simple.

When you consider the cornucopia of problems that have to be overcome to clone a human being, you realize that only a lunatic would attempt this. Enter the lunatics. On 26 December 2002, a religious cult calling themselves the Raelians announced the birth of the first human clone after several failed implantations. The baby girl, named Eve, was reported to have been born by cesarean section and to weigh 7 pounds, and the press ate it up. Have no fear though: according to Rael, the leader of the enterprising band who boasts of having 50,000 members in 85 countries, this is all part of the natural history of mankind, since we are all the product of a much larger cloning experiment. Rael (former French journalist Claude Vorilhorn) insists that all humans are actually the descendants of cloned alien scientists who were visiting Earth. Nope, nothing to worry about there.

The Raelians started a project that they call Clonaid which operates under the name of another company that they will not reveal the name of for fear that they will be investigated or persecuted. In reality, Clonaid is a registered company in the Bahamas, so this must be a double super secret subsidiary, so secret that it put out a press release under the name BioFusion Tech announcing more clones in South Korea, although the South Korean government found no wrongdoing by the company. In fact, they found no evidence that any human cloning had taken place.

To say that this group is deliberately obfuscating their organization's real activities is like saying that Paris Hilton is a little vain. After several promises to allow outside scientists to test the children and mothers to prove that they have indeed cloned a human, the Raelians have failed to provide even a stitch of evidence, prompting many to claim that they are just trying to bilk naïve people to part with their money. They claim that they did not want to risk the children being taken by the state, because a series of lawsuits had been filed against them for child endangerment. This of course sparked further speculation that the cloned children had developed severe health complications or had died, ignoring the fact that the showmanship of the cult leaders was really just a bunch of crap. The idea that these idiots were actually doing high-level science and were overcoming the developmental challenges that all the other leading scientists in the world working on animal cloning couldn't is completely ridiculous, and speculation to the contrary is really a waste of time. And if you thought it couldn't get any weirder, several members of the cult posed for a pictorial in the October 2004 issue of *Playboy*.

Another equally dramatic team of whatever you want to call them, Drs Severino Antinori and Panayiotis Zavos, also claim to have cloned humans, and they too have failed to provide any evidence. Antinori is best known as the man who implanted a fertilized egg in the uterus of a 63-year-old woman in 1994, allowing her to give birth. Just about every fertility doctor you talk to will tell you that this is irresponsible behavior, not least because the mother and child were deliberately put in danger, but also because the child will likely lose its mother before the age of 18. In 2004, Dr Antinori claimed that three cloned children had been born in procedures that he had 'advised' on. The press covered it again, but alas, no evidence has been produced. To date, the Raelians claim to have cloned over 13 humans and plan to continue implanting cloned embryos. All of the babies are said to be doing well and show no signs of the health issues associated with the cloning process in other animals. All these groups have really succeeded in doing is giving opponents of embryonic stem cell research and therapeutic cloning extra firepower to attack legitimate science. The Raelians are now under federal investigation and, believe it or not, many of the people who invested money in the group now suspect that they were defrauded. So a few irresponsible assholes have given conservative charlatans heavy weaponry and bilked a few suckers..

In the end, the activities of conservative groups and the lunatic fringe have derailed funding of legitimate science and sparked fear in the American public, who have shown up to the gunfight unarmed – without the ability to tell the difference between science and bullshit.

All the king's horses

An interesting idea cropped up in fall of 2004 that aimed to bridge the moral divide between embryonic stem cell research advocates and their opponents – define when an *in vitro* fertilized embryo is technically dead for implantation purposes and use only those for research. Don Landry and Howard Zucker, both scientists at Columbia University, published their idea in the *Journal of Clinical Investigation* and presented it to the President's Council on Bioethics. Basically, they call for a definition of embryo death akin to the definitions used for organ donation purposes. They noted that 60% of frozen *in vitro* fertilized embryos fail to start dividing within 24 hours when thawed and are thus not implanted. However, there is research that implies that not all of these embryos are abnormal and that it might be possible for them to be activated and for stem cells to be harvested from them.

Similar to organ donation, where organs can be legally harvested with permission of the family at the point of brain death, stem cells could be harvested from embryos at the point where they are determined not to be useful for reproductive purposes. The legal definition of brain death, established in 1981, was controversial, but widely accepted. At the point when a person for all intents and purposes was not going to be able to recover from massive injuries, usually caused by a severe accident, their parts could be used to save or improve the lives of others. Some embryonic stem cell opponents agreed that if markers could be developed that accurately predict the viability of embryos then this would be acceptable.

The concept creates an interesting series of slippery slopes. First, while not doing so overtly, it might allow opponents to get their way in calling a ball of cells a human being: if you are calling it dead, then it must have been alive at some point. They are going to continue to do this anyway, so I am not so worried. Brain or no brain, blood or no blood, heart or no heart, they say it is alive. The reality is that some conservatives seem dangerously close to the 'every sperm is sacred' argument. This 'the world is flat' mentality will take time to extinguish. I prefer the argument that since these embryos have no nerve cells, and therefore no brain at all, they are not people, but nothing more than human cells in a undifferentiated vegetative state. They have no consciousness. This regularly brings up the 'How do you know?' argument, which I equate to those who swear that their plants have feelings. The entire matter is just another way for abortion to be brought back to the bargaining table. Why give in to that? Legal abortion is overwhelmingly favored in most countries and by most Americans, so why engage in the debate over when life starts with people who willingly manipulate science to suit their ideological whims?

Second, the reason why these embryos are not developing properly is not understood. There could be something fundamentally different between the cells in these embryos and cells from viable embryos that makes them less than ideal for research or future therapies. Landry and Zucker point out that if you withdraw stem cells from these embryos and inject them into an egg lacking a nucleus that they will begin to divide again, so they should be fine.

Third, there is no molecular marker to predict the division between embryonic death and life, and Landry and Zucker don't really propose a way to find one. This is problematic because you would have to do human embryo experiments to determine the value of any proposed marker.

Finally, the proposal fails the common sense test. If their proposal were accepted, then it presumes that the healthy embryos would remain frozen indefinitely or until such time that they are no longer 'healthy.' Do we therefore *have* to implant or freeze all *in vitro* fertilized embryos to preserve life? And since the freezing process kills some of them and reduces viability in others, do we have a new moral argument against freezing embryos? It is good to think creatively when engaging the self-righteous, especially when doing so underlines their hypocritical views. Regardless, the philosophical argument and acceptance of the principle could be an important shift in the opposition.

On the very same day that Landry and Zucker presented their idea to the President's Council on Bioethics, one of the council members, Bill Hurlbut, a Stanford bioethicist who has openly opposed embryonic stem cell research, presented an idea of his own. He thought it would be good to create clusters of cells that divide like stem cells and have all the properties of stem cells, but since they don't have the capacity to become a fully organized embryo, wouldn't be considered human life. He went on to explain that if you remove or mutate a gene from the embryo that is vital to its later development then the entity would not have the ability to develop to adulthood and thus could be used to harvest stem cells. This proposal is a simple play on the subtleties of the word *viability*. If it can't grow into an adult then it is never viable; thus research can be done on it. The reality is that you would be creating mutant embryos that would die. Further, the stem cells would need further genetic manipulation to correct the mutation or add back the gene. He calls the idea Altered Nuclear Transfer or ANT, which is an appropriate name for the impractical 'high hopes' notion that we can just make the cells up or avoid killing an embryo by making it unviable in the first place.

It is a lovely attack on the idea of embryonic viability and brings up all sorts of largely mental masturbation notions on when we call something viable, alive or even an embryo. But at the end of the day, I think Dr Hurlbut is humping the leg of the larger political debate. He seems to presume that the early embryo could be considered a human being if it is genetically intact. This is absurd, as there are many mutations that cause diseases and early death that do not make the embryos any less human, and manipulation of a single gene would require thousands of women to donate eggs for research just to validate his whimsical notion. The truly interesting thing about Dr Hurlbut's testimony was that he claimed that scientists were enthusiastic about the idea and were interested in pursuing it. I called around and asked a few stem cell scientists about the idea, and they laughed. Not only was not a single one of them interested in investing any of their resources in pursuing ANT, none of them thought that it had any practical scientific merit. (Dr Hurlbut is a bioethicist, not a working scientist, so this is not surprising.) If Dr Hurlbut was a post-doctoral fellow and proposed this research in a grant application, he would likely get pitifully low scores from his peers, yet somehow because he is has risen in the ranks of bioethics, he is actually taken seriously. He even briefed Republican Senate staff on it.

In the end, I think this esteemed member of the President's Council on Bioethics is deluded, and while few stem cell researchers will say it publicly, many have shared the same sentiment with me. Not a single member of the council questioned the cumbersome and convoluted gyrations of Dr Hurlbut's public testimony and several of them embraced the idea, displaying how completely out of touch they are with the scientific rigor of the field. This is not surprising considering that there was not a single active embryonic stem cell researcher on the council when he presented it.

One of the biggest problems with the President's Council on Bioethics is that the membership is stacked with people who already agree with the President's stance on controversial issues. While the members are generally bright, and often articulate and accomplished, there seems little point in setting up an advisory committee that will give you a predetermined answer rather than a realistic view of a particular field. Mr Bush has actually politicized science more than any other modern president. In fact, one of the Council's members, Dr Elizabeth Blackburn, was removed from the committee after she complained that the agenda and conclusions of the meetings and reports they were putting out were not reflective of the accepted realities of the scientific fields that they were charged with analyzing. To give you some perspective, Dr Blackburn is widely considered one of the most respected and accomplished scientists in the US, is a member of the National Academy of Science, and is short listed for receiving the Nobel Prize. She specifically took issue with conclusions and inferences in the reports they were putting out. For example, the Council concluded that the goal of studying aging is to extend human life indefinitely, but that the consequence of extending life is that lifetime fertility is decreased. Neither of these statements is true. In fact, both statements are rather ridiculous. Curing or treating age-related disease to extend health and thus lifespan is radically different from seeking the fountain of youth or striding towards immortality. She also took issue with statements in one of their reports on stem cells.

That report is careful not to state the current limits of adult stem cell research, including the fact that blood stem cells are currently the best characterized and that other adult stem cell sources are still quite tenuous. It does not include information on the lack of ability to freeze or grow adult stem cells well and it most certainly does not emphasize the fact that reports of adult stem cells being transformed into cells from organs that they were not derived from are controversial because they have generally not been replicated by independent scientists and often represent experimental artifacts rather than a reliable technology. In other words, these results are preliminary and incomplete. Reading the Council's report leaves one with false impressions that reflect politics rather than science.

So what are proponents of embryonic stem cell research to do? Stalling, lying and obfuscation on the part of the government have frustrated the American public and the scientific community. While it is true that therapeutic cloning as a legitimate medical treatment is pie in the sky for now, there is a rational research design with well-defined goals, including ethical limitations on the work. Opponents of this type of research have been very successful at putting up roadblocks that have slowed research progress. No money – no work. But stopping modern biology is like stopping a freight train. With an impasse at the US federal level, private money has flowed, but not enough to make up for the lack of federal funds. Disgusted with the current state of affairs, some enterprising Californians took funding of this research into their own hands. On the November 2004 election ballot there appeared state proposition 71, which provides $3 billion to stem cell research over a 10-year period. The voters passed the initiative 59% to 41%. The state's constitution is now amended so that stem cell research must be funded at a maximum of $350 million a year, and the California Institute for Regenerative Medicine has been established. In addition, reproductive cloning is banned in the state of California, and an oversight office is established to regulate the activities of the new institute. In essence, California just eclipsed the federal government in stem cell funding. To be sure this legislation is extraordinary. Is it extraordinary in its stupidity though?

Economically, California is not doing well. The state has a history of passing legislation that is socially progressive but economically foolish. Despite its good intentions, the initiative might fall into that category. First, $350 million a year is no small chunk of change. To raise that money, the state is going to sell bonds. When one figures the interest, some back of the envelope calculating will bring the cost of the initiative up to $6 billion over 30 years: that's $200 million dollars a year. That's not a lot of money for the country's richest and most populous state, but as I mentioned, the state has not been doing too well. A lot of bonds have been issued from the state's General Fund over the years. In fact, its bond debt grew from $30 billion in 2004 to approximately $50 billion in 2005. California basically wrote a charitable check that it can't afford.

Proponents have argued that this is a sound investment for the state because the cost of healthcare annually is some $110 billion and that this research could lead to cures that reduce those costs. But we all know from experience that new therapies are very expensive and that healthcare costs have done nothing but rise. This is ignored time and time again by people who argue that research will lead to therapies that reduce healthcare costs. There have also been arguments that money from patents could pay for a significant amount of the bond. However, there is no provision that guarantees that the people of California would see a single dime of any proceeds from intellectual property that comes from the research they fund. Finally, there is the argument that it will create a lot of jobs and economic growth for California. Does anyone really believe that the Californian economy will be helped by creating jobs to support scientific endeavor? This argument has been made by a lot of people and is just bunk. Granted, this is an unprecedented amount of spending, and it will undoubtedly create a flow of scientists and technical staff to the region, but it won't create enough jobs for people who already live there to make a dent in the unemployment rate, which hovers around 6%.

Why am I so negative about this initiative? The fact is I am not. I think it's a whopping amount of money to support what will surely result in important scientific and health breakthroughs. I am negative on the rhetoric. There is a simple reason why the idea is a good one: because it is the right thing to do in the face of a federal government that has screwed its people out of an opportunity. Because research and discovery are good. Because exploration of new biomedical frontiers is good. Was it done foolishly? Perhaps. Did it show inordinate favoritism toward a single area of research? Probably. But what if this research leads to a cure for diabetes? It's quite a gamble, and I hope it pays off and that I am completely wrong.

In addition to California's efforts, biotech, philanthropic and foundation donations have ponied somewhere around $350–400 million dollars for basic research into stem cells and therapeutic cloning. Other states, including Connecticut, Illinois and New Jersey, have set aside a total of $260 million for stem cell research and are seeking more. Newer initiatives in North Carolina, New York, Maryland, Missouri and Pennsylvania are seeking to bring stem cell research to their states. Even Mr Bush's own state of Texas is getting in on the act. Most of these efforts make no distinction between stem cell research and therapeutic cloning, but a few states have outlawed reproductive cloning.

In contrast, the governor of Massachusetts, Mitt Romney, has stated that he is against the creation of or destruction of embryos for research purposes and vetoed legislation that would fund embryonic stem cell research. The legislature overrode his veto. This would have been enough to put his head on the block in the 2006 election, as the democratic base of his state has shown little tolerance of rhetoric that sounds anything like the agenda of the Bush administration. Romney, wisely, did not run again. There has been much speculation that Romney will run for President in 2008, but with doubts that he could ever carry his own state in a Presidential election, most seriously doubt he will win his party's nomination. A few states have outlawed any form of cloning, although the overreaching element that binds all of them is that none of them have what scientists would call 'strong' research programs.

Sweet and sour

President Bush delivered a stunning State of the Union address in 2005, stating:

> Because a society is measured by how it treats the weak and vulnerable, we must strive to build a culture of life. Medical research can help us reach that goal, by developing treatments and cures that save lives and help people overcome disabilities, and I thank Congress for doubling the funding of the National Institutes of Health. To build a culture of life, we must also ensure that scientific advances always serve human dignity, not take advantage of some lives for the benefit of others. We should all be able to agree on some clear standards. I will work with Congress to ensure that human embryos are not created for experimentation or grown for body

parts, and that human life is never bought and sold as a commodity. America will continue to lead the world in medical research that is ambitious, aggressive, and always ethical.

He used fear of fictional baby factories and monkey boy rhetoric to make it look like he was objectively protecting the people from rogue scientists – scientists who he thanked Congress for increasing the funding for. It should be noted that a week later he presented a budget that had the lowest increase in funding for the NIH since 1970. The increase was well below the inflation rate, so it was essentially a cut in funding – the first cut in NIH funding in 35 years. Rather than recognize that he had made a mistake in forming his policy back in 2001, he continued the lie, backing it up with fear tactics.

In his 2006 State of the Union he took it a step further, stating:

A hopeful society has institutions of science and medicine that do not cut ethical corners, and that recognize the matchless value of every life. Tonight I ask you to pass legislation to prohibit the most egregious abuses of medical science: human cloning in all its forms, creating or implanting embryos for experiments, creating human–animal hybrids, and buying, selling, or patenting human embryos. Human life is a gift from our Creator – and that gift should never be discarded, devalued, or put up for sale.

He started off implying that scientists are acting unethically and then called for bans on a combination of crazy ideas that no one is proposing and legitimate research. How anyone could frame this as anything but a deliberate misunderstanding and obfuscation of science is beyond me.

In May of 2005, the House of Representatives passed a bill that would lift the President's ban on federal funding of research using embryonic stem cell lines by a vote of 238–194. With over 200 representatives cosponsoring the bill, it seemed like a no-brainer, but the leadership refused to allow an up or down vote on the bill the year before. In fact, the House leadership had to be strong-armed into considering it by one of their own. Republican Representative Mike Castle had to threaten to hold up legislation if the leadership refused to bring it to a vote. In the days before the vote, the President went on a full public relations campaign against it, stating, 'This bill would take us across a critical ethical line by creating new incentives for the ongoing destruction of emerging human life.' He also held a rare press conference at the White House where he stood with families who had adopted frozen embryos, so-called snow-flake babies. He deliberately misled the public, giving the impression that this was a viable option for the 400,000 embryos currently thought to be in frozen storage, even though only around 100 babies have been born through this adoption service.

During the debate before the vote, the later disgraced House majority leader Tom Delay piled on the rhetoric, stating that 'An embryo is a person,' and that 'This bill tramples on the moral convictions of an awful lot of people who don't want their tax dollars to be spent for killing innocent human life.' Throughout the debate Republicans stated over and over again that adult stem cell therapy held far more promise than embryonic stem cell research, even though this is the opposite opinion of the scientific community. Suddenly, everyone was an expert, using the same talking points. Along with the stem cell bill, the House passed another bill that specifically sets aside money for research on blood cells from umbilical cords, setting up the theme for the debate in the Senate, where 'viable alternatives' to embryonic stem cell research are presented by conservatives who don't understand the science they are arguing for. Cord blood has been shown to be a useful avenue for treating blood diseases like anemia, but there is no indication that it could be used to treat other diseases. The House failed to pass a bill criminalizing reproductive cloning because Republicans would not consider the bill unless it also banned therapeutic cloning.

After the bill passed the House, stem cell supporters in the Senate called for a rapid vote on the bill, confident that they had the votes to pass it. This was a significant show of force, because it included Republicans like Orrin Hatch of Utah and Arlen Spector from Pennsylvania, representing a rare breaking of ranks with the President on a hot button issue. The Senate majority leader, Dr Bill Frist, a former transplant surgeon, initially remained silent as Republicans did their best to devise a strategy to spare the President public embarrassment. In the mean time, a poll was released that indicated that the majority of Republican voters actually favored a lift of the ban, enraging many Republican congressmen, who were upset that they were not informed that the poll was being conducted in their district by the Republican pollster who initiated it.

In a matter of weeks it became clear what the Republican strategy would be – confuse the issue. Senator Frist indicated that there would be as many as six or seven stem cell bills considered and asked Democrats to consider voting on all of them as a package. The idea was that the President could then sign some of the bills that did not lift the ban so he would not appear to be anti-embryonic stem cell research. Democrats rejected the offer out of hand, calling for a separate vote on each of the bills. At this point, Senator Frist was considering simply not considering any stem cell bills, but did take the time to consider bills to repeal the estate tax and prohibit liability lawsuits against gun manufacturers.

At the end of July, it was suggested that if the bill was not considered, it could be attached to other bills as an amendment, which could force the President to veto an energy or defense spending bill. This forced Senator Frist to buckle in the face of public pressure and say that he would bring the bills to a vote in September when the Senate returned from summer recess. He also changed his mind and indicated that he would actually vote for the bill despite the fact that he felt that there were significant ethical issues that were not addressed in it. September came and went and the bill was not brought to a vote. He promised again in December, but the bill was not brought up for a vote until May 2006. The President then used his first and only veto to shoot down a bill that the overwhelming majority of people favor. Not exactly a man of the people or for the people.

While Republicans were considering how they would derail stem cell research efforts, the National Academy of Sciences released their guidelines for embryonic stem cell research, which included clear rules to prevent reproductive cloning, prevent the exploitation of women for eggs, and ensure informed consent. The guidelines were a thoughtful set of straightforward regulations for the field, indicating that the scientific community is fully capable of ethically regulating itself without the intervention of Congress.

States' efforts have certainly more than made up for the cash that the government would have likely dedicated to stem cell research, but the President's policy has disrupted the normal distribution of research efforts and caused a bias against smaller efforts at universities outside of the major state initiatives. This is causing a distinct research disadvantage for many institutes and states, leading to researchers having to migrate to the money, and forcing them to uproot their families. They have also become vilified in the eyes of the citizenry who still look to the President for moral guidance.

Like most things scientific, with stem cells and cloning the devil is in the detail, and misunderstanding of those details or flat-out paranoia are more common than not in society. This is to be expected, in the US especially. Our education system stinks. Most people have a rudimentary understanding of science at best and believe that having a deep education makes you elitist, not lucky. Even our President presents himself as having no intellectual curiosity so that he comes off as a 'regular guy'. The leader of the free world doesn't show pride in his Yale and Harvard education, and make no mistake about it, the public digs that.

While conservative Republicans, including the President, displayed their awesome contempt for the scientific community and tried to deliberately mislead the public, scientists continue to make rapid advances towards using embryonic stem cells for therapy. At the same time, Democrats in the House of Representatives brought up and passed the stem cell bill within the first 100 hours of the new congress. Guess who hasn't figured out that he is on the wrong side of the issue yet? The stem cell controversy will not be going away, and neither will intolerance on both sides of the argument.

3

there's sex ...

We are one uptight nation when it comes to sex. There's no question: when it comes to discussing sex in any way we invariably go right to, and get stuck at, rhetoric about there being too much sex in our society and how children are being negatively influenced by rampant images and filthy language that concern sex. There is a lot of material to cover regarding sex, but of most concern to us is that science and health are often at the center of the controversy, and people don't seem to be listening to the facts.

The issue is not about what science has and has not told us about sex. Despite the fact that we are all the product of the intimate work of our parents' genitals (yes, even IVF kids), and that we are all prone to sexual desire, the guts of the sex controversy rotate around morality, choice and an inevitable denial of our biology. Most of the flap concerns choice, or when and how people have sex and who does it. Like most controversial topics that concern scientific findings, there is a negative element in society that will cherry-pick information to suit its moral agenda. Rather than weigh all available information and draw logical conclusions that can be used to enlighten us about sex and sexuality, these moralists are so convinced they are right that they filter information and obfuscate statistics to recklessly promote an agenda without serious consideration of personal privacy, civil rights and legitimate scientific data. Unfortunately, there is often an undercurrent of religion, ignorance and fear, or some combination of the three that forms the views of those who will remain right despite being incorrect.

When we talk about sex, we have to be clear about what we mean. First there is the genetic meaning of sex; simply put, those who carry two X chromosomes are female and those who carry an X chromosome and one Y chromosome are male. Next there is gender: in other words, is one living as a man or a woman? After that comes sexual orientation, and here we will go with the traditional Western Judeo-Christian definition of heterosexual, homosexual and bisexual. Finally, we have sex drive and mate choice within a given sexual orientation.

In this chapter we are going to talk about actual sex: you know, doing it. Along with this come all the trappings of reproduction, including contraception, pregnancy, abortion and disease. See, isn't it fun to discuss sex in such sterile terms? Actually, that brings up the next topic: sex education. While fascinating, I don't think there is any true controversy around much of our normal sexual differentiation and what makes a baby become male vs. female. Questions along the lines of why men have nipples if they don't breast feed are not what we are interested in; we're after the places where science, health and culture collide. (For the record, men have nipples because the fetus develops along a female developmental pathway for the first six weeks of life. Nipples are set out before other sexual differentiation begins.)

A hard study

Scientists have been interested in the biological and social underpinnings of sex for as long as there have been scientists. What is interesting is that controversy always stirs, no matter what they are studying. Today we have genetic, biochemical, neurobiological, endocrinology, psychological, epidemiological and social scientists that all approach different aspects of sex in unique ways. What they give us is a broad picture of biology that often runs counter to widely held beliefs about sex. This is where we will focus.

'Is the love of man and woman merely an animal function? Are spiritual ideals of mating, of fidelity and chastity no more than irrational and sentimental nonsense? Have our conventions and moralities – and what we've always held to be simple decency – been outmoded by findings of modern science?' These are all common questions that society wrestles with routinely. But you would probably be surprised to learn that they are quoted from a *Reader's Digest* article that appeared over 50 years ago in response to the publication of the first large-scale studies of human sexuality by Dr Alfred Kinsey and colleagues. Dr Kinsey actually published two volumes on human sexuality. The first, on male sexuality, entitled *Sexual Behavior in the Human Male* came out in 1948 and became an instant bestseller. Five years later, the second study, on female sexuality – aptly titled *Sexual Behavior in the Human Female* – also became a bestseller and set off a storm of controversy. The studies were not just controversial because they were the first of their kind; the actual results knocked the public on their asses by showing that the moral code that people were preaching was often diametrically opposed to their actual behavior. The Kinsey report was the first scientific study to point out that there was a great sexual hypocrisy in American culture and many claim that the venom spewed in this pioneer's direction led to his failed health and early death.

For all of the hullabaloo they caused, the volumes themselves make a rather boring casual read. They are dry, cold and extremely analytical. The facts are presented with as much scientific objectivity as one could possibly muster in the face of such shocking results. The public was simultaneously aghast, intrigued and shocked. Many sexual activities and preferences were discussed seriously for the first time. They revealed that almost all men masturbated and that most women also masturbated, despite being taught that it could lead to disease and depravity. Not only that, most people fantasized while doing it. Sexual taboos like cunnilingus, fellatio and premarital sex were shown to be common. The reports revealed that sex frequency declined with age, but that older people still enjoy and engage in sex and masturbation. It revealed that men and women peak sexually at different ages and that homosexual acts and fantasies were far more common than most had suspected.

The Kinsey reports are certainly the most important scientific research ever done on human sexuality, not because they were so large and comprehensive, but because of the lasting influence they had on society and the field of sex research. Almost immediately after the initial report on male sexuality was released, conservatives attacked the research as sloppy and biased. Even today, a quick search of the web will find scores of websites dedicated to tearing apart Dr Kinsey's research methods and calling his results lies. They also attack him personally with accusations of pedophilia, coercive homosexuality and perversion – anything to discredit a man who was a true pioneer in an area that was previously shrouded in secret. These sites never mention the fact that Kinsey succeeded in a seemingly impossible task and revealed more truths about human sexuality in his work than any other sex researcher has since that time. He probed society's taboos about sexuality when people still believed that women generally remained virgins until marriage, that men were either heterosexual or homosexual, that only men masturbated, and that infidelity was rare, to name just some of the startling results. Even today, interviewing as many people as his group did with such detailed personal questions would be viewed as an overly ambitious task.

What most people don't realize is that Kinsey was not trained as a psychologist or even a social scientist. He was trained as a zoologist, and worked cataloging and categorizing millions of specimens of a relatively obscure type of wasp for most of his career. Yet, his methods were as good as and often better than any social research done at that time. What is really remarkable is that while many of his numbers have proven to be overestimations of human sexual behavior as a result of a sampling bias, the trends that he reported have generally held up over time even as society evolved and people became more open about their sexuality.

Fifty years on, the tremendous cultural opposition to open discussion of sex and sexuality still makes it difficult for researchers to acquire grant money to adequately study issues in this area. This is despite the fact that it is now crystal clear that sex in all its connotations has a dramatic effect on health and society as a whole. The reason for the opposition is that scientists are often seen as pushing a cultural agenda through their research. This is a mistake that Kinsey made in his career at certain points, but that most sex researchers do not make today. The problem is that any scientific results that point in a direction not in line with conservative thinking is immediately deemed invalid, a fabrication or not credible. Some vocal religious conservatives feel that there is a vast conspiracy of sinners who want porn on television and shtupping on the front lawn, and they feel they need to fight it. Like most things these days in which religious extremists get involved, so does Congress. This is because the Republican Party has a large contingent of people who put the Bible before the health of their constituents.

Since 2000, there have been several incidents where select congressmen in the House of Representatives have requested a list of federal grants relating to sex topics. They have also openly questioned the use of federal dollars to study sex and questioned the amount of money spent on finding cures for diseases like AIDS. The Bush administration has also put internal pressure on the Department of Health and Human Services not to fund grants for sex research. Conservatives have argued that they don't like federal dollars to go to these issues through the National Institutes of Health and the Substance Abuse and Mental Health Administration.

One particularly ugly incident in 2003 involved a memo sent from a staffer from the House Subcommittee on Criminal Justice, Drug Policy and Human Resources. The memo specifically raised concerns with the NIH about two grants having to do with controlling the spread of sexually transmitted diseases. One was for research on how to prevent the spread of sexually transmitted disease by controlling it in prostitutes and the other on sex and drug abuse in the transgender community. The memo expressed concerns that studies intended to protect the health of sex workers 'seek to legitimize the commercial exploitation of women.' They went so far as to ask who headed the study section that evaluated the grant and how it fared against other grants. The same staffer authored another memo asking about another study on preventing the spread of HIV in the gay community and demanded that the NIH provide him with a list of HIV prevention studies. Soon after the memo was sent, just by coincidence, the researchers received site visits from the NIH and other federal agencies they were funded by to make sure there were no 'irregularities.' Now, you have to understand that it is highly unusual for a congressman to get involved with a specific grant from the NIH, but it is even more unusual for a grantee to receive site visits from multiple agencies to look into the matter. Unfortunately, judging the merits of scientific enquiry by congressmen has become common since the Bush administration came into power. This heavy-handed oversight by Congress is completely counter to the basic principles of science funding: that grants are awarded based on merit, not ideology.

It has become so bad that program staff at the NIH are now warning grant applicants to avoid terms like 'needle exchange,' 'transgender,' 'sex worker,' and 'condom' in their grant applications, so that non-scientists who have no ability to evaluate scientific grants, are unfamiliar with how to better the healthcare system and are in the pockets of extremist groups don't get their hair up over them. This is all happening under an administration that has cut funding to programs that would get sex workers off the street, reduce drug trafficking and reduce poverty. The hypocrisy of these conservative politicians is nothing short of shocking. The Republican staffer who sent the memo now works at a high level in Senator Tom Coburn's office. In other words, this guy was not fired, but was promoted to a senior position. Unbelievable! Despite the Draconian efforts of scientifically inept congressmen and their staff, the research continues, just at a slower pace than it would in a more progressive society, and at the cost of human life.

Makin' babies

While most people are devoid of the intellectual curiosity necessary to wrap their heads around complex issues like stem cells or alternative fuel research, when it comes to what other people do in their bedrooms or in child bearing it seems that everyone has an opinion. This is particularly evident in the ever-present abortion debate, the sentiments of which are echoed in the public's opinion of every aspect of conceiving and raising a child. Despite such stringent opinions, most have a hazy memory of past debates or are helplessly unarmed with knowledge. When technology gets involved in making babies, science has plenty to say, but not on when and where such technologies should be used. Interestingly, despite wide public belief to the contrary, scientists and doctors have never claimed or wanted to have the last word on making children.

Take for example discussions of artificially selecting the sex of a child. It is not surprising that everyone, whether they understand the technology and purpose of sex selection or not, has some conclusive statement to make about the moral use or science fiction-based misuse of selection techniques. Let's get it straight before we go on.

There are two basic ways to select the sex of children before they are born without utilizing abortion. The first is to fertilize embryos *in vitro*, let them develop for a little while, suck out a cell and then diagnose it for the presence of two X chromosomes or one X and one Y. This is a fairly expensive procedure that was developed to rule out devastating diseases in embryos before they were implanted. For example, if you knew that you carried a genetic factor that would doom your child to a horrible early death, you might choose to screen embryos before implantation to ensure that you have a healthy child. If that disease marker is on the X chromosome, and only causes disease in men, then you might select for only female embryos or male embryos that did not have the bad copy of the X chromosome. Most people don't have a problem with this activity, and geneticists are coming up with tests for more and more diseases that can be screened this way.

The other way to screen for the sex of your children is by running sperm from the father to be through a machine that can separate sperm with an X chromosome from sperm with a Y chromosome. It does this by looking at the overall DNA content of each sperm with a laser. The X chromosome is many times larger than the Y, so the difference is easily detectable. This technique is about 75% effective for creating male embryos in the clinic and 90% effective for creating female embryos. As you have probably already guessed, companies have started offering both of these techniques to couples who do not have any particular health concern for their child, but just want to choose to have a little girl or boy. This is where scientists step away from the picture and greed and vanity step in. The techniques were developed for health purposes, but are being applied to make predetermined family ratios.

The use of preimplantation genetic diagnosis for non-medical purposes has been outlawed in many countries, including Canada, Australia, Britain, France, Germany, and India, based on moral and ethical concerns that include the ever-present slippery slope argument. People are afraid that if we start choosing children based on sex, what is to stop us from choosing their height potential, IQ, or artistic talent? The answer is simple – science. We are not likely to be in a position to choose these things in the near future because of the technical limitations of the tests. There are too many factors that go into determining intelligence and stature.

Personally, I find the concerns to be correct, but completely hypocritical. As a society we are OK with people going to the plastic surgeon in record numbers to give us looks that we were biologically never meant to have, but we draw the line at choosing characteristics for our children. Both attitudes seem to be steeped in vanity and both are dangerous precedents because they deny natural processes.

I think the line is arbitrary in a society that encourages the indiscriminate use of fertility drugs for couples, resulting in quadruplets and even septuplets, but which does not hold doctors accountable for the negligent use of these drugs. Multiple births of this size are dangerous for the mother and children, but we call them miracles and give the families free minivans and home makeovers rather than look critically at how such a screw-up could have happened. Society does not have a problem with an infertile couple using IVF, but does have a problem with them deciding that they want a little boy or girl. Society draws the line at certain steps in the process, but not on the process as a whole. Opinions on the use of reproductive technology range quite a bit. Polls have, however, shown that strict evangelical Christians generally believe that *in vitro* fertilization is a bad thing if they are told that excess embryos are made, but that it is a positive thing if the question is framed as a family-building technology.

One recent public opinion poll on the use of reproductive technologies by Johns Hopkins Genetics and Public Policy Center showed that only 60% of Americans think that sex selection of embryos for non-medical reasons is a bad thing and only 65–70% thought that selection for intelligence and strength was a bad thing. Truly shocking was that around 25% thought that selecting against having a child with a fatal disease was immoral. Amongst those who assigned maximum moral worth to an embryo in a dish, about half still thought it was OK to select for an embryo that would not have a fatal disease. In other words, the embryo in a dish has equal life to an adult unless it will be really sick and die later. More telling is the background of respondents in recent surveys where 80% of evangelical Christians thought that reproductive technologies would lead to us treating children like products. This number is similar to the 73% of those without a college degree who felt the same way. Those of other faiths and with college degrees ranged around 50–60%. So negative feelings about reproductive technologies correlate somewhat with being an evangelical Christian and/or uneducated and about one third think that it is OK to select a child based on fictional genetic tests for intelligence.

Despite the wide-ranging feelings about all types of assisted reproductive technologies, and the often hypocritical views of what is done with spare *in vitro* fertilized embryos, the practice continues to grow in popularity. This is partially due to moral acceptance and decreased costs, but is also due to the fact that couples are often waiting longer to have children, sometimes well into their forties.

While *in vitro* fertilization is considered quite safe, there has yet to be a large long-term study on its effects on the health of children. This is not due to a lack of trying. At least one grant application that I know of for such a study was rejected at high levels within the Department of Health and Human Services despite the fact that it scored very highly during the peer reviewing process of grant approval. Not surprisingly, the scientists involved do not want to be identified for fear of retribution. There are several possible reasons for this. Assisted reproductive technology is big business in the US, and if some evidence popped up that there should be concerns about some techniques used by some doctors or companies, it could sour the public against using it. Also, heavy-handed oversight of grant funding by Bush administration officials within HHS has led to investigations and internal warnings about grants that were being approved. There are a growing number of anecdotal stories that indicate that there has been a clampdown on any studies having to do with assisted reproduction. This is of course very bad, but not surprising knowing what we know of the administration's approach to scientific findings that do not suit its ideological agenda.

Unfortunately, there is some evidence that there is a need to study the long-term effects of assisted reproductive technologies. Scientists studying a rare disease called Beckwith–Wiedemann syndrome discovered that approximately 4.6% of their patients were conceived using *in vitro* fertilization. Beckwith–Wiedemann syndrome results in a variety of birth defects, including several types of cancer and excessive growth of some tissues. Nationally only 0.8% of live births are from *in vitro* fertilization, so there is a six-fold increase in the number of *in vitro* fertilization babies who carry this disease. This is not evidence to determine cause, but it does indicate that further investigation is warranted. The very need to resort to *in vitro* fertilization to conceive a child could indicate that there is a genetic or metabolic problem. Perhaps some of these problems could lead to Beckwith–Wiedemann syndrome or other diseases in their children. After all, *in vitro* fertilization is really just forcing the issue of reproduction on a biological system that for some reason does not want to conceive. In most cases, forcing pregnancy through new technology is not detrimental, but without looking we will never know if disease could be prevented in some IVF children by adjusting the technique if some practices are shown to even remotely increase the chance of a particular disease.

While the use of assisted reproductive technology for becoming pregnant still represents a small minority of births, there are some concerns about more common procedures on the road to having children. For example, a recent study documented that cesarian section is now the most common procedure performed in hospitals. Approximately 30% of all births in the US are by cesarian section, costing an estimated $14.6 billion in total hospital charges and representing over 90% of overnight hospital stays. It is hard to come up with a precise number, but many medical professionals claim that most c-sections are not really medically necessary. The problem comes in the subjective definition of 'medically necessary.' There are reports that they reduce some complications during birth and thus reduce the number of law suits in the US, but that is not a good definition of necessary. The statistics are particularly disturbing because the cost of healthcare is skyrocketing, hospitals are often losing money and there is now evidence that women who deliver their first child by c-section are at higher risk of complications with subsequent pregnancies, including more trouble getting pregnant again and longer delays between children. Overall, cesarian section is a very safe procedure and women should not stop having them for medical purposes, but those considering unnecessary surgical procedures should always question what the precise reason is. In light of the overly expensive healthcare system in the US, doctors and insurance companies should be seriously asking whether medically unnecessary procedures should be performed and covered by health insurance.

Technology comes in many flavors, not all of them tasty. One of the most understudied areas in science is how environmental factors created by new technologies affect health. Even worse is the dearth of research being done on how environmental factors interact with or affect genetic factors in health. This is particularly important for pregnant women. The question is: Which synthetic substances cause disease and birth defects?

There are some alarming trends in men's health that are suspected to be associated with environmental exposure to chemicals. For example, there has been a constant and sometimes dramatic rise in the incidence of testicular cancer depending upon where you live. Also, the sperm count of American men has been steadily declining for the last century, while the incidence of deformed sperm has increased dramatically over that same period. There has also been a dramatic increase in autism, hyperactivity disorders, attention deficit disorders and other mental disabilities that tend to affect men more than women. Finally, and most disturbing to many men, is the fact that certain types of genital birth defects have been increasing. While the increase in disease can be partially attributed to increased diagnosis, when it comes to Mr Happy and his merry men, changes in medical diagnosis don't account for the trends. It's sad, but true, that men would rather have their genital tackle working than their brains, but all of these trends should really be looked at far more carefully than they have been. Collectively, the evidence seems to indicate that there are elements in the environment that are doing bad things to people. No surprise there. Hormones that regulate genital development are an obvious first place to look. So scientists have been asking what environmental factors can affect normal hormone balance.

Today, there are serious concerns about a class of chemical additives that are biologically active, meaning that they affect biological function by interfering with or mimicking the action of natural molecules in the body. Of particular concern are chemicals that affect hormone pathways like estrogen and testosterone signaling. In fact, the chemicals are collectively called estrogenics because they mimic the effects of estrogen or interfere with estrogen signaling. Some of these chemicals, called phthalates, have been positively linked to male genital deformity and are the likely culprit for a host of other medical problems. Estrogenics are used in many modern products, including pesticides, wood finishes, hairspray perfumes and plastics. Ever get a good whiff of new car smell? Well, that is caused by phthalates leaching out of a car's dashboard and other plastic parts. This is why the smell is more pungent on a hot day than in the cold.

One of the better known estrogenics is a pesticide called DDT, which was banned from use in the US in 1972 because of the reported effects on animal development and the potential risk to humans, since this class of chemical is associated with birth defects and some mental disabilities. In animals it caused sex switching in some amphibians and was highly toxic to birds and fish. When Rachel Carson revealed her theory on the impact of DDT on the environment in her book *Silent Spring*, she was widely cast as irresponsible by conservatives eager to protect the interests of industry. To say that the widespread use of DDT was a major environmental screw-up is like saying that diet might be contributing to our nation's weight problems. At the same time, DDT is responsible for saving untold numbers of lives when it was used to control the mosquito population in areas where malaria is prevalent. So like most things in science, there is often another side of the story.

In the late 1990s, the European Union voted to ban some estrogenic chemicals from baby toys because there was research that indicated that the chemicals might be the cause of health problems in children and later in life. This is not a case of the evil chemical companies doing bad things, since the chemicals are actually used to make plastics softer so as to make children's products safer. While the evidence was sometimes quite weak, the EU erred on the side of caution. In retrospect, the EU missed a major point. It is widely recognized that the most critical time in child development occurs in the first trimester of pregnancy and that the fetus in general is very sensitive to changes in hormone levels. So banning these substances in children's toys will have no effect on the incidence of birth defects. They could potentially prevent some developmental disabilities that can occur later in life, but without further research into how environmental factors cause birth defects we will never know. Similar attempts have been made in the US, most recently in the city of San Francisco, but have all failed because here we err on the side of business rather than people, even in the face of scientific evidence that indicates a potential danger to the public. Nice!

Since these estrogenic chemicals are affecting estrogen signaling, one would think they would affect women as well. This is not necessarily true, though. Estrogen signaling in women will normally vary, so it is reasonable to believe that abnormal estrogen signaling would be masked because women's bodies are used to fluctuations in the signal. There is one place, however, where you might see it: prepubescent girls, who do not normally have a high or fluctuating estrogen level. And this is precisely what is seen. The puberty age for girls has been steadily lowering over the last couple of decades. Normally, estrogen levels spike around puberty and girls start to develop secondary sex characteristics like breasts and pubic hair. In one study, Puerto Rican girls who had developed breasts prematurely were shown to have on average a seven times higher concentration of a specific estrogenic compound than a control group. Data in mice also indicate that estrogenics have an effect on sexual development. So, we have a whole range of evidence that there is something to this environmental effect on sexual health, but no smoking gun. What do we do?

The government established the Center for the Evaluation of Risks to Human Reproduction which put out a draft report

on some phthlates and their risk to human reproduction. It
found a range of concerns over their effect on human reproduction, but did not recommend that they be removed from
items like food packaging, even though they had concerns
that they could damage sperm producing cells in the testi-
cles and result in low sperm count and damaged sperm in
humans. How is it that more Americans are not deeply concerned about demasculinizing birth defects in their male
children and premature sexual development in girls? Part of
the problem is that there are not that many people studying
the effects of environmental chemicals on gene expression
and health in animal models. The truth is, nothing really
happens on these issues when you have a government that
will bend over backwards to protect the interests of industry
over the health of the citizenry. The trends are alarming and
certainly indicate an urgent need for immediate research
into the issue. So what are the press, politicians, religious
leaders and general public in the US interested in? That's
right, erectile dysfunction drugs.

It's getting harder every day

When Viagra hit the market in 1998, a few women's health
advocates mistakenly cried foul at the notion that drug companies were creating drugs to deal with male sexual dysfunction but not women's. Further, they reiterated the claims
that our sexist culture still has not come up with a male birth
control pill because men are not willing to take responsibility
for contraception. In fact, Viagra was discovered by accident
in a clinical trial for a drug intended to treat angina. Men in
the trial were reporting a surprising and sometimes awkward
side-effect that in some cases improved some aspects of their
personal relationships. Bingo! The fortuitous birth of a multi-
billion dollar industry.

Birth control pills are a different story. It is far more simple
and has everything to do with biology. Since women ovulate
once a month and the hormones that regulate ovulation are

well known, it was relatively easy to control it: simply sustain the hormone that prevents ovulation. The ultimate problem with regulating sperm production, on the other hand, is that men make sperm pretty much all the time. Worse, the key driving hormone is testosterone, which, not coincidentally, is also responsible for sex drive. So, if you stop testosterone, you stop the desire to have sex. Granted, this would be a *really* effective form of birth control, but it misses the point of contraception development though. Comparing stopping ovulation to stopping sperm production is like comparing shutting the windows when it is raining outside to actually stopping it from raining outside.

I don't know about you, but I don't know many men that wouldn't opt to take a pill to prevent unintentional fatherhood responsibilities. If we count on men being as shallow as we think they are (a safe bet), you would have men bragging about being on the pill. Also, the accusations of no effort going into male contraceptives miss the minor detail that there has actually been a lot of work done to develop a male pill. It really is that hard to develop drugs that stop natural processes. The work that has been done to refine female birth control is essentially refining the original invention. The hardest part of drug development, or any discovery for that matter, is coming up with the first working version. Are there loads of areas in society where women are still getting the short end of the stick? You bet. Birth control development is just not one of them. Which brings us back to Viagra.

Since drug makers knew that Viagra works by blocking a certain receptor protein, it was far easier for subsequent drug companies to develop drugs that work the same way and grab a piece of the pie. As long as the drug compounds do not look too similar to Viagra on the molecular level, they are able to get into the boner biz by blocking the same receptor. So it didn't take long for Cialis and Levitra to hit the market. People wanted these drugs and it would have been irresponsible to company investors if drug companies didn't chase down their own versions of Viagra.

In addition, there were some initial clinical trials to see if Viagra could work in women. Since the drug works by increasing blood flow to the groin, in theory it could work to stimulate blood flow to the clitoris and enhance the sexual pleasure of some women who have problems with physical excitation but not desire. A common misconception is that Viagra can increase libido. It doesn't. Here we run into another fundamental difference between men and women – sexual dysfunction. This can also be framed as sex drive, but for the purpose of discussing treatment it is best to stick to medical convention: what is wrong and why?

For most men sexual dysfunction in the form of erectile dysfunction comes without an effect on the desire to have sex. Simply put, they don't get their F-on because they can't get their F-on, not because they don't want to get their F-on. So, for many men, drugs like Viagra are real life-changers. Many doctors are concerned that the drug allows men to ignore the underlying cause of the problem though. This is of particular concern because erectile dysfunction can be a side-effect of serious blood pressure problems, heart disease and a host of other medical conditions. It is also common in men who have their prostate removed during cancer therapy.

In 2005, it was reported that a few men who were taking Viagra regularly had started to go blind, sparking predictable jokes on the late-night talk show circuit and calls by the reactionaries to pull the drug from the shelf until there are further studies. But these cases were few compared with the number of people taking Viagra. What was reported far less frequently is that many of the underlying causes of impotence can lead to blindness if left untreated and that these men were not usually being effectively treated for the causes of their dysfunction, just for the dysfunction itself. It is also true that many Viagra prescriptions were being handed out for recreational use so that men could have sustained erections. This again ignores the fundamental danger of taking any prescription medication, but erectile dysfunction drug sales have skyrocketed because of recreational use.

It should be pointed out that not all erectile dysfunction is caused by serious health concerns. It can also come with the normal decline in testosterone in men as they age, the so-called male menopause. Unfortunately, not a lot is really known about male menopause. For a long time, it wasn't even recognized as a part of normal aging. It had, however, been known for some time that testosterone levels fall with age. Besides keeping people sexually active, which can have very positive effects on mental health, the Viagra craze has made it OK for men to talk about a problem that before would have devastating psychological effects and is commonly hidden from doctors and significant others. It also brings up the hypocrisy of a nation that feels comfortable with television ads for erectile dysfunction drugs during prime time television, but has a problem with a television show that discusses sex or an ad for contraception during the same time slot.

One aspect of sex drive that most people don't discuss or even think about is that there is natural variation in sex drive in men and women, and in most cases sex drive declines as we get older. There is no question that self-image and relationship dissatisfaction are more likely to affect a woman's sexual desire than male sexual desire. This is not to say that women do not lose sex drive as a result of physical problems – they do. They can also lose the ability to become aroused as a result of physical problems – it is just not the most common issue with female sexual dysfunction. Sex drive is different from sexual function.

All of these issues have made it very hard to make drugs to treat female sexual dysfunction. Frankly, it is not completely clear what to treat. This is echoed in the draft FDA guidelines for companies interested in producing drugs to treat female sexual dysfunction. First released in 2000, the draft guidelines drew quite a bit of criticism even though they were based on the medical guidelines from the 4th edition of the *Diagnostic and Statistical Manual of Mental Disorders*. Simply put, the guidelines identify four possible problems in female sexual dysfunction: (1) decreased desire to have sex; (2) difficulty having orgasm; (3) decreased arousal; and (4) experiencing pain during intercourse. The way FDA guidelines work is that a company seeking drug approval must pick which aspect of the disorder they are aiming to treat. The guidelines caused quite a stir because they failed to recognize that sexual desire and arousal in the form of natural lubrication and increased blood flow to the genitals are intimately linked and that, experiencing pain during intercourse can be linked to arousal and that desire and pain as well as arousal and pain can be linked. There was so much controversy over the guidelines that the FDA never released a final version.

So, if we consider that a common sexual problem in women is desire and that desire is intimately linked to the other physical signs of female sexual dysfunction in women, then the best bet is to treat desire. Right? The problem with treating sexual desire like it is a disorder in need of a magic pill is that it ignores the fact that desire is not strictly a physical problem. Often it is the result of mental trouble, from dissatisfaction with a relationship, poor self-esteem, trauma or some other underlying issue. Also, some women naturally have lower desire than others. As with men confronting their lack of ability to get it up, confronting the underlying problem for women is hard to do because of cultural stigma.

Now, before you freak out at my generalization that sexual dysfunction in men is a physical problem and sexual dysfunction in women is a mental problem, realize that they are exactly that, generalizations. There is tremendous overlap in the spectra of causes for sexual dysfunction in men and women. But the generalization that desire and arousal in women are highly linked and should be approached that way when thinking about treating sexual dysfunction should not be controversial. Desire issues in men can also be linked to psychological and social problems, but are more often linked to physical ones than is the case for women.

Since there are no approved pharmaceutical treatments for women, doctors have been prescribing testosterone off-label to treat decreased sexual desire. That's right, testosterone! Both men and women normally have both estrogen and testosterone in their body, just in differing amounts. Both of these hormones decline with age in men and women. We know that testosterone can regulate sex drive, mood, bone density and muscle growth in men, but it can also have the same affects on women. Testosterone therapy in women has been reported to have a very positive effect on sexual desire. This is particularly important for women who experience a drop in desire as a result of menopause, because testosterone also has the potential to increase bone density and muscle. (Loss of both comes along for the ride after menopause.) It has also been shown to be beneficial in women who have their ovaries removed, since testosterone levels in these women drop significantly as well.

After the success of Viagra, the drug industry quickly recognized the potential in the opposite sex and started developing many patches, creams and pills either containing testosterone or designed to increase testosterone levels in women. Other drugs are designed to increase blood flow to the genitals without regulating testosterone. But it will take a bit of public education to get many women to agree to increase their testosterone levels for fear of becoming 'manly.' Also, there are some safety concerns in the public as well as the medical community associated with long-term artificial regulation of hormone levels in women. A few years ago, it was discovered that long-term treatment of post-menopausal women with estrogen can lead to an increased risk of developing heart disease.

Since the idea behind the treatment of women is that desire, not arousal, is the issue, you will not wind up with a female Viagra that you take just before you think you might want sex. It will have to be a sustained delivery. In a clinical trial for one testosterone treatment, 36% of the women who received the placebo indicated that they would like to continue the treatment before it was revealed that they had not actually been treated with any medication. Placebo effect is common in all fields of clinical research, but that is an absolutely staggering number and an overwhelming indication of a mental component to their problem. In my view it is also a damning indictment of a culturally warped society where many women still feel sexually repressed.

Related to this issue is orgasm. It is undoubtedly a major difference between the male and female sexual experience, but it is also an important indicator of the cultural divide. The numbers are telling: approximately 25% of women say they almost always have an orgasm, 45% say that they sometimes do, and approximately 20% say they rarely have an orgasm. The final extreme is the 10% of women who claim they never have an orgasm under any circumstances. There is a lot of variability here, and much of it is culturally, socially and experience based. Also, different surveys have come up with different numbers, but on the whole these are pretty sound figures.

What purpose does the female orgasm serve? The initial answer is obvious: to make a woman feel good so she will want to engage in sex. The whole truth is not nearly as obvious as for the male orgasm, which is pretty much a requirement for procreation so it is quite easy to see why making this feel good could lead to an evolutionary advantage. There have been all sorts of theories as to what the evolutionary advantage (if there is one) to the female orgasm is. Some research has suggested that, during orgasm, women suck more sperm into them, but that has been pretty well refuted. Some of the other theories sound good, while others are quite weak. I prefer the 'Who gives a shit?' theory. Whether the female orgasm is a fortuitous left-over or encourages lots of sex and therefore procreation really doesn't matter; it is mostly intellectual masturbation to ponder it.

What is of interest is why some women have them and others don't. Again, for many the answer is cultural and psychological. There have been some studies that suggest that there is a strong genetic tie for ability to orgasm. These studies did not look at DNA, but instead considered twins. Basically, they asked female twins about their orgasms and then did some statistical corrections for environment. They concluded that there was very strong evidence that some of the ability was inherited in identical twins. This is pretty lame evidence, but you have to understand that these sorts of studies are commonly used to see if there is justification for conducting more detailed (and expensive) DNA-based studies. There is certainly reason to believe that there are variable biological underpinnings to the big O, but in our uptight society geneticists can rarely, if ever, get funding to look at anything even remotely related to sex. So the orgasm gene may very well be there. Until we have some serious cultural change, the unfortunates who do not have it are out of luck unless some fortuitous drug is found in some heartburn trial down the road.

Contraception misconception

Preventing pregnancy is one of the most bizarre issues in America today. The rhetoric is full of ridiculous hypocrisy and religion-based stupidity about human reproduction. It is common to hear people say that poor people should stop having children that they can't afford to pay for and also say that they are against the distribution of birth control products or the teaching of sex education in school. This is a fundamental misunderstanding of our biology, and simplistic solutions like teaching abstinence will only exacerbate the problem.

The issue is well highlighted by the hype over emergency contraception. Commonly referred to as Plan B or the Morning After Pill, it is intended for use when condoms fail, when people do not use contraception, and perhaps most importantly in rape cases to prevent unwanted pregnancy. In essence it is a large dose of normal birth control medication and has been shown to be safe and effective if taken within a few days of intercourse. Considering that there are approximately 3 million unintended pregnancies in the US each year and approximately 1 million abortions, it would seem that most people, especially anti-abortion advocates, would want this medication to be widely available. But it also seems that logic is not one of the strengths of these groups, who mounted strong opposition to the FDA approving the drug for over-the-counter sales. 'Why?' you ask. Because there is a small chance that it could affect the lining of the womb and cause an implanted embryo to abort.

These groups are often confusing the Morning After Pill with the abortion pill, RU-486, which does cause an implanted embryo to abort by triggering the lining of the uterus to shed as it does normally once a month. The truth is that all of the scientific data to date shows that the Morning After Pill acts by preventing or delaying eggs from being released from the ovaries. The problem is that science cannot rule out that in some cases it also has some effect on implantation, but to date there is no evidence for this and since the Morning After Pill is not effective if a woman has already ovulated, the chance of it working to induce abortion is in fact very low. This has not deterred these groups, as they use a different definition of pregnancy than the medical community. They define pregnancy as the time when sperm and egg fuse: fertilization. The medical community defines it as when an embryo implants. So there is a difference in opinion between those with a medical education and anti-abortion advocates who generally have a biblical education.

A high percentage of fertilized embryos never implant and are washed out when menstruation begins. One would think that if the anti-abortionists truly believe that these embryos are the equivalent of children that they would hold a funeral at every menstruation, try to recover the dead embryo for a memorial ceremony, or at the very least say a prayer while they are wrapping the sanitary napkin or tampon in toilet paper to be discarded along with coffee grinds and last night's creamed corn.

In December 2003, an FDA advisory panel voted 23 to 4 to recommend that Plan B be granted over-the-counter status, but the FDA took no action. There are exceedingly few examples of cases where the FDA has not gone with the advice of its own expert panels. These cases almost always revolve around public pressure rather than true medical concern or confounding scientific data. Take, for example, silicone breast implants.

After it was revealed that there was no solid evidence of a link between silicone breast implants and autoimmune disease, the FDA should have probably followed the advice of its advisory panel and allowed their sale again. However, the FDA sent the manufacturers back to do more studies to make doubly sure. Fair enough: they are not depriving people of an important medication. But in the case of emergency contraception it is estimated that close to 51,000 abortions were averted in 2000 through its use and there have been no serious adverse side-effects reported. Since emergency contraception was only available with a doctor's prescription (except for six states), that number will likely skyrocket and it is estimated that close to 700,000 abortions could be averted. That would be a serious victory for anti-abortion advocates and for women who would not have to go through an invasive procedure or take RU-486, which can induce significant discomfort. Since there is no question that emergency contraception is safe and would save women from having to undergo an invasive procedure, why did it take the FDA so long to rule on it?

In 2005, Senators Clinton and Murray held up the Senate confirmation of acting FDA director Lester Crawford for his failure to resolve the Plan B issue. In short, they wanted him to prove that he had the ability as a leader to resist pressure from outside groups when issuing rulings on controversial drugs and devices. The point was that medicine, science and health should automatically trump rhetoric. As soon as Crawford announced that the FDA would make a ruling on it, Senators Clinton and Murray allowed the vote to happen and he was approved by the full senate. A few months later, the FDA announced that it would delay its final decision on Plan B indefinitely because there was no way to guarantee that it would not be used by women under the age of 17 without parental consent. You know – requirements like showing a driver's license to buy it, as for cigarettes and alcohol. Crawford deliberately misled the US Senate and proved that he is nothing more than a political lackey. The director of the FDA's office of women's health actually resigned in disgust over the decision. Her name is Susan Wood. She should be given a medal for this act of protest. (Since the first printing of this book she has received many awards and is widely regarded as a hero for women's rights.) Incidentally, Lester Crawford resigned soon after Dr Wood and only a few months after his Senate confirmation. Essentially, he wasted the time of the Senate and didn't hold up his end of the deal. In late 2006, he pleaded guilty to charges of lying to the Department of Justice about his financial interests in companies that the FDA regulated and broke conflict of interest laws when he was acting Commissioner of the FDA. Susan Wood? Hero. Lester Crawford? Criminal.

In November 2005, the Government Accountability Office released a troubling report on the decision process used to evaluate the emergency contraception pill 'Plan B' for over-the-counter status. They found that the directors of the offices that reviewed the application, who would normally have been responsible for signing the Plan B action letter, disagreed with the decision and did not sign the not-approvable letter for Plan B. They also found that the FDA's high-level politically appointed management was more involved in the review of Plan B than in those of other over-the-counter applications and that the decision to reject the application was made before the scientific review was completed. Finally, they found that the rationale for the Acting Director's decision was different from the 67 other prescription to over-the-counter decisions made between 1994 and 2004, including unprecedented consideration of the cognitive development of adolescents who might take the drug.

There was a final twist in the rejection of over-the-counter status for Plan B. One of the three anti-abortion doctors that the administration had placed on the committee, David Hager, had written a minority report of the decision at the request of someone (he wouldn't say who) at the FDA. In a speech he gave at a Christian College he stated 'God has used me to stand in the breach for the cause of the kingdom.' He went on to state that he 'argued it from a scientific perspective. And God took that information and He used it through this minority report to influence a decision. You don't have to wave your bible to have an effect as a Christian in the public arena.'

The other argument against allowing emergency contraception to be sold over the counter is that it will encourage risky sexual behavior. However, there is no evidence to indicate that. What is really the issue here is the lack of availability of regular contraceptives and a failure to teach young adults about the benefits of their use. Condoms in particular are the best protection against pregnancy and contracting a sexually transmitted disease if you are having sex. I can't believe I have to point out that not having sex is an effective means of not getting pregnant, but that is where our debate has turned. It's like saying that you won't get fat, if you decide not to eat. Used properly, condoms are 98–100% effective in preventing pregnancy and sexually transmitted diseases. That is the key: used properly. A lot of the data that refers to the condom failure rate includes people who did not use a condom every time or used the condom improperly. Even in those cases, people who identified condom use as their main form of contraception only had a failure rate of 15%. These are not complicated devices: if kids can work an iPod, they should be able to manage a condom. But kids today are not being taught how effective condoms are or how to use them. Instead, they are learning about abstinence.

Those who preach abstinence have a good point. It is next to impossible to get pregnant or contract a sexually transmitted disease if you are not having sexual contact with anyone else. Similarly, it is impossible to get fat if you refrain from eating. It is a total denial of our biological drive to have sex ... or eat. That is the point that the administration has made through the Department of Health and Human Services websites and a large campaign in schools to teach abstinence. In fact, the administration had spent close to a billion dollars between 2000 and 2005 trying to persuade American teens that abstinence is the way to go and that condoms are not as effective as people think. In 1988 about 1 in 50 high schools had abstinence-only programs. That number has skyrocketed, so that today about one third of American High Schools teach abstinence only. The administration has also given tens of millions of dollars to religious groups that urge children to sign an abstinence pledge. Several of these programs have come under fire for pushing religion hand in hand with responsible behavior and providing factually wrong information on sexually transmitted diseases and pregnancy. By 2006, the situation had not changed and a Government Accountability Office report showed that the majority of abstinence-only education programs had still not been vetted for scientific accuracy and, as you might expect, religious undertones are still found in many of the programs. The use of federal funding for the spread of religion is expressly forbidden by law.

One such program, called the Silver Ring Thing, has received significant taxpayer dollars to urge teens to take an abstinence pledge and wear a silver ring as a symbol of the pledge. They put on multimedia shows including loud music, comedic sketches and lights designed to appeal to teens. The group claimed that approximately 75% of those who attend their shows signed the pledge. This would seem an enormous victory for the program, but there is a catch. The show preaches that condoms are not effective in preventing pregnancy and contraction of sexually transmitted diseases.

In the eyes of the US government, this is exactly what they are supposed to do. For the Silver Ring Thing and other programs like it to receive federal funding, they can only talk about the failure rate of condoms and other forms of contraception, not their success rates or how to use them in the event that a teenager *is* going to have sex.

The facts don't back this policy and again highlight how the Bush administration has put religious ideology ahead of the health and wellbeing of Americans. For example, in the largest study ever performed on teenage sex practice, researchers interviewed 20,000 adolescents and found that abstinence programs will on average delay teens from having sex for 18 months. However, when they do have sex, they are 30% less likely to use a condom and before they have vaginal sex are more likely to engage in risky sexual acts like anal sex. The study also found that these teens are less likely to get tested for sexually transmitted diseases and to be secretive about being sexually active despite the fact that they are just as likely to contract a sexually transmitted disease as teens who do not take the virginity pledge. Sandwich those facts in with the statistic that 88% of teens who take abstinence pledges before marriage break the pledge, and you have an ugly picture.

Teens are getting information that teaches them that sex is taboo and that using a condom is a waste of time because it will not protect them. The leader of the Silver Ring Thing actually said that if his daughter were to break the pledge, he would not want her using a condom because it would give her a false sense of security. Many of these organizations, including the Silver Ring Thing, are now being investigated for using tax dollars for 'inherently religious' activities. The ACLU filed a suit against the group for using tax dollars to teach religion.

These religious groups will be spreading their message on not using condoms to Africa in the coming years. Good thing there isn't an AIDS or TB problem there. Otherwise that would be like the US exporting death squads. I just hope they can be prevented from entering countries where they could really cause harm. Let's be 100% clear about the logic. Do not teach the benefits of condom use because they have a failure rate around 15% when improperly used, but spend millions of dollars promoting programs that result in an 88% failure rate, risky sex activity, do not affect the STD rate in teens, and teach them that protective contraception is not something you should be using if you are going to have sex. Not telling teens to use a condom and how to use it because of a failure rate is the equivalent of telling someone not to wear a motorcycle helmet because it won't always protect you when you crash.

There is a true war here: religious ideology vs. health. Since the investigations and law suits, the Silver Ring Thing has removed statements defining its religious purpose from its website. Statements like 'The mission can only be achieved by offering a personal relationship with Jesus Christ as the best way to live a sexually pure life.' They also removed two steps from their twelve step follow-up program, including step two, which called for abstinence Bible study, and step four, which tells teens that God has a plan for his or her life, and a plan for his or her sexuality. Can you say guilty? In 2005, the government finally cut off funding for the Silver Ring Thing, but only after *60 Minutes* and a deluge of print stories brought it into the non-Christian fundamentalist light. The bizarre reason was not that they push irresponsible health policies, but because they fused it with religion. The government suspended their funding and in 2006, and the Silver Ring Thing announced that it was not going to seek further federal funding to spread its message. Truckloads more of these irresponsible groups remain under the cloak, and the government is signing checks left and right to fund them without policing their activities.

The consequences of not providing adequate sex education are severe. In a 2005 survey of 16–18 year olds, it was revealed that 52% of them thought that chlamydia only affected women, 31% thought that sexually transmitted diseases could be caught from a toilet seat, 54% did not know that emergency contraception can be used up to 72 hours after sex to prevent pregnancy, and that approximately 30% of young men and 20% of young women did not know that sexually transmitted diseases could be transmitted by having oral sex. This indicates that all the progress society has made toward educating the public about sexually transmitted disease has been lost in a matter of years by dismantling the education system. This puts teens in immediate danger, since we know that many of them are going to have sex. We know they are learning about sex from movies, television and magazines, not their parents and teachers.

It is estimated that the average woman would have 12–15 children in her lifetime without the use of contraception. But there have been dramatic increases in the number of pharmacists that are refusing to dispense birth control or emergency contraception to patients who have a legitimate prescription in dozens of states. Often, these pharmacists are located in towns where they are the only means for people to get prescription drugs. With no alternative means of acquiring contraception, women in these towns are being subjected to the moral objections of their pharmacist despite the fact that they are trying to acquire an FDA-approved, doctor-prescribed medication. These pharmacists are not just located in mom and pop pharmacies either; they are in national chains like CVS and Walgreens. The individual cases are disturbing: from women being refused emergency contraception after being raped to refusing mothers with multiple children the pill to prevent them from having more children or who wish to space out their pregnancies as part of their family planning. There have been refusals to transfer prescriptions to other pharmacies and clients have been lectured on morality and had their prescriptions voided. All of this, it seems, is perfectly legal in most states.

While it is expected that a pharmacist's first duty is to ensure the health and safety of his clients, this cannot be achieved when one overlays moral judgment on the health decisions made by women in consultation with their doctors. This is particularly disturbing in cases where women use birth control as part of the normal treatment for other disorders, like endometriosis. Over 30 states have had legislation introduced that would make it legal for pharmacists to refuse to dispense medication on moral grounds. If passed, many of these laws would allow pharmacists to refuse to prescribe HIV medication to gay people, cancer drugs to people who smoke or medications associated with stem cell therapies.

There are extremists on both sides of this argument, but it would seem that a middle ground can be struck where a pharmacist can refuse to dispense some medications on moral grounds if and only if there are legitimate alternatives available within a reasonable distance. It should most certainly be illegal for a pharmacist to refuse to transfer or destroy a legitimate prescription on moral grounds, and there should be stiff penalties for pharmacists who refuse to provide medication when there is no alternative provided. That penalty should include loss of license and jail time when it is determined that a moral judgment endangered someone's health.

The irresponsible actions of some pharmacists and the ludicrous policies of the Bush administration are not the only serious issues in contraception. There are several Republican

HIV
4 children are infected every minute
1 child dies every minute
72 adults are infected every minute
5 adults die every minute
40.3 million people living with HIV in 2005
2.2 million children are infected
25 million people have died since 1981

Source: World Health Organization

congressmen, including Senator Coburn from Oklahoma, who happens to have an MD, who have been pressuring the FDA to require a label on condoms that indicate that they do not provide protection against passing or contracting several sexually transmitted diseases, especially Human Papilloma Virus (HPV), which causes genital warts and most cases of cervical cancer. Their efforts are based on a law signed in 2000 by President Clinton that requires that the FDA provide language for manufacturers to put on condom labels that indicates their effectiveness for preventing sexually transmitted diseases. On the surface, this sounds like a good idea. But for most diseases we do not have very accurate numbers on the efficacy because research on disease prevention with condoms is under-funded. So the FDA basically could not comply. Conservatives have seized upon this as evidence that condoms don't work, but that is absolutely false. It's a revolving door of rhetoric. They do not want condom effectiveness studied with federal funds, but complain that there are no accurate numbers on their efficacy in preventing disease.

Internally at the FDA there was pressure only to report failure rates, not the effectiveness, which is in line with the Bush administration's policies. To say there was an internal stalemate is understating the frustration felt within the department. Several of the congressmen would like the labels to read that they are not effective in preventing STDs, which is simply not true. They are an excellent preventive measure for all sexually transmitted diseases and most of the failure rates are linked to lack of consistent use or improper use. There are of course some diseases that can be spread without intercourse, so the allowable heavy petting that many Christian teens now think is OK can actually spread disease. Also, condoms just cover the shaft of the penis, so herpes and HPV can infect and be spread through the skin that is not covered, but rather than teach proper and consistent use, kids are being taught that they are unsafe. This is absolute insanity, and will lead to the death of men and women by HIV and cervical cancer caused by HPV because they just didn't get the message. In 2006, the FDA approved the first vaccine against HPV, but not without controversy from the religious right, many of whom believe that a vaccine against a sexually transmitted disease would encourage promiscuity in teens. One leader actually said that he would be against the delivery of an HIV vaccine for the same reason. Here, the true colors of the extreme religious right come into clear focus. They are less concerned with the health of people than they are with people having sex outside of marriage. Since 70% of all cervical cancer is caused by two types of HPV, it was nothing short of astounding that they would take this stance. In the end, they were forced to soften their rhetoric. With approximately 1 in 20 high school girls estimated to be carrying chlamydia and a reported rise in most sexually transmitted diseases in the US since the President took office, it seems that the abstinence education policy is pretty ineffective.

4

... and then there's sex

Men and women are different. (The crowd cheers.) Seriously though: people still don't get this. The differences go way beyond the equipment between your legs and the painfully obvious differences that are expounded on at great length in excruciating romantic comedies starring Meg Ryan. Pointing out these differences or their possibility can get you into trouble if you don't word them carefully. I'll try not to do that too much, as the politically correct vernacular can sometimes get in the way of clarity and the occasional argument that is worth having.

The root of the differences between men and women is not cultural, but genetic. A single chromosomal difference means everything. Swap an X for a Y and the entire experience for that person changes, despite there being 45 other chromosomes that have quite a bit to say about biology. It is the chromosomal, and thus biological, difference that has led to the culturally based differential treatment of men and women that has developed through the millennia. While debunking some of the common falsehoods about the true difference between men and women leveled the playing field in many important areas, it has also resulted in the pendulum swinging the other way for some. The women's movement was an important step towards cultural equality between men and women. But the strides that have been made have led some to forget or deny that there are some significant biological differences between men and women. Some have lost touch with the fact that the inborn differences can be useful, and are often very important. The point is that choosing to treat people equally and judging them on personal merit rather than genetic make-up does not preclude the existence of innate biological differences that are important to understanding health, culture and fundamental biology.

Birds vs. bees
Take drugs for example. For a long time women were not included in clinical trials for drugs. In fact, the FDA ruled in

1977 that women should not be included in clinical trials for new medications when men could be used. This was basically a precautionary backlash policy resulting from the controversy stirred by Thalidomide, a drug prescribed to women starting in the mid-1950s to battle morning sickness. It was later found that Thalidomide caused severe birth defects of the limbs, ears and internal organs. The logic was well-intentioned – women should not be involved in clinical trials because of unknown risks to a fetus in women who are unaware that they had become pregnant. It was assumed that men and women would react the same to drugs, effectively ignoring the possibility that genetic and physiological differences between men and women could affect the comparative safety and effectiveness of medications. As it turns out, this logic was largely correct and the great majority of medications work equally well in men and women and dosage has been easy to adjust to women's physiology.

There are three basic places where a drug could cause different effects in women and men: absorption, processing and activity. The first two can change the relative concentration of a given drug in the bloodstream, causing either too low or too high a concentration. Many of the differences in processing of drugs have been linked to the activity of liver enzymes. About half of all drugs are metabolized through a single liver enzyme that causes women to clear some drugs faster than men.

The relative activity of a drug is hard to pin down because so many factors can affect it once in the body. The ugly truth is that clinical trials are often required to see if there is a true difference in activity of a drug in men and women. These trials will often be quite complicated because of confounding processing and absorption differences. One notable study on the topic showed that of 300 FDA-approved drugs on the market, only 11 were found to have processing differences greater than 40%, and none of those differences resulted in different clinical outcomes. Thus no labels were changed, but that is not always the case.

With the rate of new medications coming out and a better understanding of how sex can influence drug response, the FDA switched its policy on including women in clinical trials in 1993. The move was made in part through pressure from proponents of the Women's Health Initiative, an ambitious project started in 1991 in which over 160,000 women were recruited to take part in clinical trials to test the effects of different therapies on women's health. This was, in essence, the largest study into the relative effectiveness of drugs on men and women without outwardly doing comparative studies.

As a result of the initiative, some notable exceptions to the general rule that drugs work equally well in men and women have surfaced. For example, a study that showed that taking aspirin regularly in small doses could significantly reduce the risk of heart attack in men was assumed to have similar affects in women. However, in 2005, over a decade after the study in men was published, it was shown that aspirin had no affect on heart attack risk in women. It did, however, reveal a reduced risk of stroke. Curiously, aspirin does not affect stroke risk in men.

Also of note are studies showing that several antibiotics and antihistamines can cause heart rhythm problems in women, but not as often in men. Opiates are reportedly more effective in women than men and there is a variation in a single gene that, believe it or not, makes opiates more effective in red-headed women but not red-headed men. Other drugs like anticoagulants cause excessive bleeding more often in women as well.

Drug companies have fought the requirement of including more than one sex in their clinical trials because factors like hormonal changes with a woman's menstruation period can affect drug efficacy and could confound a study. This is not just big pharma not wanting do the studies to ensure that their drug is safe and effective. While it is true that it is just not possible, for monetary and logistical purposes, to do clinical trials in every possible population, the more significant reason really is scientific. If you mix the clinical trial, you could confound your data. Thus, they do the study in the population that is easiest to find volunteers for: white men. But this really shouldn't be the end of the story. Like all things in medicine, a happy middle ground can be found where smaller pilot studies on women can be done and reported. If there is no detectable difference between men and women, then there shouldn't be a need to do a large-scale study on both sexes. In other words, keeping clinical trials simple is necessary for detecting safety and effectiveness, but pilot studies early in the process could save lives and produce valuable information.

While many have seen the Women's Health Initiative as being important for finding information about women's health, others have been quite critical of its implementation and true value. The price tag and slow pace of the study is somewhat disturbing when we consider what we are getting out of it. The aspirin study, for example, cost $30 million, took a decade to complete and concluded that its beneficial for stroke in women, not heart attacks. Most women still ask the same question 'Does that mean I should take it?' The answer for many doctors is still unclear.

We now have pretty convincing data that differences in effectiveness or danger for medication are a rare exception, not the norm. I think the real value of the initiative is not the results that will come from their drug studies. Aspirin is not likely to save a tremendous number of women from stroke. The true value is an increased awareness that there are fundamental differences in the metabolism of men and women and that chromosomal differences count for much more than genitals. There have not been a lot of problems associated with simplified clinical trials and most of those could be addressed by much smaller pilot studies than those on the level of the initiative.

As a side note, there has also been some misunderstanding in the public about the degree of disparity between money spent to research women's diseases versus money spent on primarily men's diseases – a big misunderstanding actually. The fact is, about twice as much money is spent on women's diseases over men's diseases, and that has been the case for decades.

The physiological differences in drug response are a small part of the overall health differences between men and women. Besides the alarming lack of ovarian cancer in men and prostate trouble in women, common diseases are also disproportionately represented between the sexes. Here we encounter one of the most confounding elements in health: environmental effects. For the most part, environmental factors like chemical exposure, cigarette smoking, and exposure to sunlight are really just triggers for genetic events. Each of them acts like a switch that causes the body to react. Sure, some environmental factors just cause gross damage and necrosis to tissue, but the majority of them just turn genes on and off.

So how do you tell the difference between an environmental factor affecting genes and a mutation in the DNA code that is causing disease? Some cases are easier to figure out than others; for example, diseases like mesothelioma that are caused by breathing in asbestos fibers primarily affect men who encounter it more often in factories and on construction sites. Other diseases, like breast cancer, can be the result of inherited mutations in genes that don't cause the disease, but greatly increase the risk of developing it, or of environmental factors that cause mutations associated with sporadic cases. There are no hard and fast rules about what causes different patterns of disease between the sexes, but environmental exposure does give us some hints. Other hints can come from the pattern of heritability. Often diseases are present only in men because the mutation that is inherited is on the X chromosome. In some cases, women get the disease only when two bad copies are inherited, but men get it when just one copy is inherited.

Most differential disease patterns are a total mystery because the factors that cause them are so complex. Mental disorders are particularly vexing, because in addition to environmental exposure and genetic factors there are cultural factors that can play an important role. Men tend to have higher rates of schizophrenia, autism, attention deficit hyperactivity disorder and drug addiction, while women tend to suffer from anorexia, clinical depression and anxiety disorders more often. Teasing out whether these differences are biological or based on social and cultural pressures is extremely difficult, and thus we know less about mental disorders than we do about just about every other disease category. Since many of these disorders show significant heritability from one generation to another and are present in all cultures, it is likely that innate genetic susceptibility plays a far greater role than was previously expected. Many researchers feel that the differences are caused or maintained by hormonal differences between men and women.

Sex hormones regulate a vast array of genetic and physiological factors, so by taking a 'where there's smoke there's fire' approach, scientists have started to uncover many of the underlying causes. We know that anxiety and depression are tightly linked and that in women these emotions can be regulated by estrogen. In men, testosterone seems to lighten the load, which explains why men do not respond to stressful situations with fear or anxiety as easily as women do. At the same time, men tend to respond to such challenges with aggression and exhibitions of violence more readily as a direct result of testosterone regulation.

So, in response to the same threat, women will often feel anxiety or fear where a man might respond with aggression or other outward behavior, like drinking alcohol. The difference between externalizing and internalizing emotions is directly related to hormonal differences and is thus more pronounced during puberty, when hormonal regulation can be erratic. Tying hormonal differences to disease has been harder to do, and some sociologists argue that these differences (and indeed the associated disorders) are manifested by learned social behaviors. While there is evidence that social conditions and learned behavior can exacerbate differences, the underlying biology is certainly there to start with.

So there are physiological differences between men and women, and these can sometimes lead to problems with the application of prescription medication and cause differences in disease susceptibility. Big woopty doo; let's have a party. What about differences under normal disease-free conditions? Do biological differences between the sexes lead to differences in ability?

One X or two?

In January 2005, Harvard University President Larry Summers got into a hell of a lot of trouble when he suggested that part of the reason why women have not made as much progress in math and science as they have in other fields could be partly due to innate differences in analytic ability. Dr Summers was lambasted by both male and female colleagues at the institution, and hauled over the coals by the press and women scientists around the country. While Dr Summers was not particularly articulate in his arguments and clearly intended to be provocative, his real fault was touching the third rail of a contentious issue without at least acknowledging the fact, if he knew it, that his idea was not new and was not supported by rigorous scientific inquiry. In case you were wondering, he is no longer President of Harvard.

The undisputed fact is that there is a large disparity in the number of men and women working in most quantitative sciences, particularly at the highest levels. However, the evidence indicates that innate differences in ability will be more pronounced on an individual to individual basis rather than between sexes. There could also be innate biological barriers to women attaining positions in the sciences that are not related to their personal ability, but to perception of them by others. Here we run into cultural differences vs. ability to perform tasks. The data is somewhat fuzzy in this area because of the complexity in assessing innate vs. acquired ability. We certainly do not expect there to be a 'math' gene or a 'can't do math' gene, and the data seem to indicate that differences would lie at the individual level, not in generalizations that can be made for men and women. In other words, the cause of innate ability is so complex that any man or woman can be extraordinary at a particular task.

Here are some of the facts. Men and women show significant anatomical differences in many areas of the brain used for cognitive function. The evidence seems to indicate that differences in brain structure and utilization could be the basis of the differences in the types of disorders they are susceptible to. The question is whether this has something to do with overall ability as well. Since there are also differences on average in their performance and attitudes towards math and science, there is a tendency by some to associate the two.

For example, male brains are about 10% larger than women's brains, even when you correct for differences in body size, but women have a higher overall percentage of gray matter, the workhorse of the brain. Also, brain activity studies have shown that men and women use different areas of their brain and differing amounts of gray matter when solving problems. Men also use more gray matter than women to solve the same problem sets. While it is tempting to put weight on these neuroanatomical factors, historically they have been overinterpreted. Alone they are rather poor predictors of normal cognitive ability. In addition to anatomical differences, there are also clear differences in performance on different types of memory task and perception. For example, men do better on tests that require one to visualize an object in three dimensions, or on tests requiring them to read a map or find a hidden clue in a complex image. Women, on the other hand, kick the crap out of men when it comes to tests of ability to read emotions on people's faces, and on language skill tests.

Some of the differences appear to show up quite early, but not all of them. Prepubescent boys and girls perform about the same in science and math. After puberty, US girls report that they have less of an affinity for math and science than boys do and they tend to perform worse on standardized math exams than boys. However, in some cultural settings girls outperform boys. There is no solid evidence that these differences reflect innate biology, but there is evidence that there are cultural influences. Even taking this into account, there are few countries where girls outperform boys in math, and in all countries studied, girls express disinterest in math or even dislike at a rate far higher than boys, regardless of test scores. The performance results on the whole do not indicate a strong advantage for boys over girls in math, but do indicate a very strong cultural influence.

Here is where it gets a little sticky. Standardized tests are about the worst predictor you can use to identify future mathematicians and scientists. The correlation between test scores and career path is weak at best. And consider that, today, women outnumber men as biology majors at the undergraduate level despite performing worse on standardized tests and are quickly catching up and sometimes surpassing men in enrollment into graduate-level biology degree programs.

I get very tired of all the social interpretations that tend to go along with these studies. The popular over-interpretations are a pile of shit and sometimes provide excuses for sexist behavior. Historical trends give the right picture in my book (which this happens to be). For centuries women were far outnumbered in law, medicine and many other fields. But that has changed. Now, the stragglers like politics and science, are starting to even out.

I know, I know: I wish there was a clear answer as well, but there isn't because, as I said, innate ability is hard to gauge over cultural influences and acquired skills. If it is there, it is negligible against a torrent of cultural biases against women. Personally, I think there has been too much emphasis on the innate ability part of the issue, but if there is an innate cognitive difference, one thing is for sure: culturally, we put more emphasis on the types of activity that men are perceived to be better at, which most certainly puts women at a disadvantage.

Are achievements in arts and science more valued in culture than someone being able to understand others' emotions or learn five languages? Definitely. And since innate ability in men and women, if it exists at any significant level, can often be overcome, what science should start to consider more seriously is if there is a biological underpinning to the cultural bias against women. In other words, are we biologically programmed to undervalue women over men, and if so, does an advanced society ignore that or correct it?

A few studies indicate that the bias against women in technical careers is strong. For example, there was a clever study where a bunch of mathematicians were asked to evaluate the merits of a math research paper (so you can imagine they were bored to tears). As one might suspect, when the paper had a female author the score was on average lower than when the exact same paper had an author who was a man. The really interesting part of that study was that women also rated the male-authored paper higher than the female-authored paper.

More recently, Princeton University students were asked to evaluate two potential candidates for a faculty position in the engineering department. Seventy five per cent of the time the students chose the candidate with a higher level of education over the candidate with more engineering experience. The catch is that when the more educated candidate was identified as a woman and the one with more experience was identified as a man, the better educated candidate only got 48% of the vote. Besides displaying predictable academic snobbery amongst Princeton undergraduates, it also showed a distinct level of sexism.

Brass tacks: there could be a biological difference in innate ability in science between men and women, but there is no solid evidence for it and if it is there, it is not a big deal. There is definitely a cultural bias against women in technical careers that still persists despite advances made in other fields. This bias could reflect a biologically based sexism, and this should most definitely be explored. All totaled, arguments for a biological basis for cognitive ability are bullshit because the evidence is weak at best. The behavioral correlates do not seem so deeply rooted that they can't be overcome by cultural change. This is fundamentally different from behavioral areas like sexual orientation, where there is absolutely no evidence that cultural change or medical treatment can overcome the innate biological drive behind sexual attraction towards the same sex or the opposite sex. To the credit of the biological community, most of the research focus has not been on trying to find biological correlates to behavioral disparity or performance, but more on the disparities between men's and women's health. This is a particularly vexing issue, and since the launch of the Women's Health Initiative we have come a long way towards understanding what these correlates are.

Naturally gay

The evidence that sexual orientation is biological ranges in believability quite a bit, but collectively the evidence is strong. First there are animal studies: many anti-gay activists have claimed that homosexual behavior is seen only in humans and that since humans are the only animals that display rational choice in their behavior, homosexuality must be a choice. The only problem with this argument is that it is wrong. Homosexual behavior has been observed in birds, beetles, orangutans, dolphins, fruit bats, sheep and many other animals, male and female. This behavior ranges from oral sex, as observed in female Japanese macaques, who engage in cunnilingus, to complex courtship behavior and sex in birds, to anal sex, which has been observed in many species. Some have argued that homosexual behavior is used by animals to enhance their social position within a group, but others have observed that they do it simply because they like it. For example, one of the most horny primates is the bonobo, sometimes called the pygmy chimpanzee. Bonobos engage in sexual activity with many partners, 75% of which is not reproductive sex; they are thought to use sex as a substitute for aggression, as a greeting and for a great deal of pleasure including homosexual encounters between both male and females.

Some of the most important information we have about the biology of sexual orientation comes from sheep. Farmers have observed that about 8% of male sheep preferentially court and have sex with other male sheep and will ignore female sheep even when there are no male sheep present. Australian sheep farmers used to call these sheep 'shy breeders.' In 2004, a group of scientists found that the area of the sheep's brain (called the sexually dimorphic nucleus) that is responsible for controlling sexual behavior, feeding and blood pressure is twice as large in heterosexual sheep as it is in homosexual sheep. This region is also twice as small in female sheep as heterosexual male sheep. Since this difference exists in sheep before birth, what is happening is not a result of the sheep's experience as an adult, but a developmental pathway that was set before adolescence. The sheep studies were done at the US Department of Agriculture Sheep Experiment Station in Idaho. Interestingly, the USDA will not let their scientists discuss the matter with the press for fear of a backlash from people who do not think that it should be studied.

There is a controversial study that showed a similar difference in human males, which came out in 1991. The controversy on that study comes not just from advocacy groups, but from the scientific community, as it has not been confirmed and was done post mortem, so the full sexual history of the individuals is not known. However, an attempt to replicate the 1991 human results showed a statistical trend that this brain region was smaller in homosexual men, but did not find a statistically significant difference in the number of cells in this region. Also, the size difference did not quite reach the level of statistical significance. Those researchers urged people not to rush to judgment in either direction until more studies are done. It is, however, now accepted that the difference in size for this region does exist between men and women. The male sheep results seem to indicate that there might be some correlation between sexual orientation and brain anatomy.

The next step in sheep experimentation is to determine what is causing this behavioral difference. Some studies indicate that exposure to different hormone levels during fetal development could be the cause, so the next set of experiments for the sheep is to regulate hormone levels during *in utero* development and see if they can affect the sexual orientation of sheep. Those experiments are now under way, but it will take a few years of testing and observation to get to the bottom of it. What will be initially unclear after artificial regulation of hormone levels *in utero*, if indeed they result in being able to regulate sexual orientation later in life, is whether these hormone changes correspond to what is going on in humans and if it is caused by the mother's behavior or environment, genetics, or some combination of factors. Nonetheless, it will bring us one step closer to answering the question once and for all.

Sheep are not the only experimental animal system for which we have an indication that sexual orientation is biological. In 2005, a researcher working with fruit flies discovered that if you modify a gene in female fruit flies so that they express the male version of the protein, they will adopt male courtship behaviors and actively court female flies. Now you have to understand that fruit fly courtship behavior is actually quite complicated. Male flies have a complex ritual that rivals some of the activity you'll see in a bar on ladies' night. The gene involved is called *fruitless*, because it had previously been shown to affect courtship behavior in male flies when mutated. Depending upon the mutation, *fruitless* mutant flies will show indiscriminate courtship behavior, have anatomical defects in the muscle that allows them to curl their abdomen under to mate with female flies and have trouble with the courtship dance they do to lure females. That this behavior is regulated by a single master gene is remarkable in itself, but that it can be bestowed upon the opposite sex is absolutely incredible and is excellent evidence that in animals at least there is a genetic component to sexual orientation.

These are not the only examples of animals exhibiting homosexual behavior or of experiments showing that there is a biological explanation for homosexual behavior. For example, increasing testosterone levels in newborn female rats masculinizes their behavior for the rest of their lives. Similarly, lowering testosterone levels in newborn male rats causes them to mount other male rats as adults. I should warn that animal studies are often taken to mean that there will be strict correlation in humans. This is most definitely not the case. However, some of the most fundamental work on human biology is based on work done in animal systems. While it is highly unlikely that single genes will ever be found to regulate human behavior, much less sexual orientation, the point should not be lost that the animal work will inevitably point us in the right direction when studying complex human biological issues.

Many claim that human sexual orientation could not be genetically based, because, through Mendelian laws of inheritance, the trait would die out. Since homosexuals are not procreating and their genes are not being passed on to the next generation, any hypothetical gay gene should die out. This would be an accurate assessment of the situation if it were not for one pesky problem – most traits are not controlled by single master genes. This is a common misconception about heredity. Single genes are responsible for some traits. This is best demonstrated by rare diseases, where mutations of single genes can cause disease. Most behavioral traits – those that have a direct impact on reproduction in particular – are almost always complex, meaning that many genes are involved. Most common diseases fall into this category as well. Second, there are significant environmental factors that can be associated with complex traits. So a single genetic factor that contributes to variation in sexual orientation will never be a catch-all for either homosexuality or heterosexuality.

Research into human sexual orientation has progressed at a snail's pace in the US, but more recently large grants have actually been funded to look for genetic factors that impact sexual orientation. This is because the piecemeal work done previously all points to the presence of genetic factors at play. First of all, identical twin studies have revealed that if one brother is gay, that there is about a 50% chance that the other brother will be gay too. In fraternal twins the number dips to about 20%, which makes sense when you consider that on average they will carry about half of the same genetic make-up. Again, the numbers have been refuted by subsequent studies, but the trend remains. Also, there are some infamous finger length studies that have been done that show that straight men tend to have index fingers that measured tip to knuckle are shorter than their ring finger. In women and homosexual men, they tend to be the same length. Lesbians tend to have finger length ratios more similar to those of straight men. While a bit simplistic on the surface, it is a significant trend that indicates that homosexuality is at least correlated with developmental events that take place in the womb. Then there are family correlations like the fact that 8–12% of gay men have other family members that are gay. The rest of the population hovers around

3–4%. (The number of 1 in 10 that many still use actually came from the original Kinsey report and was found to be biased based on sampling error.)

There is an interesting theory based on correlations of birth order that seem to have some weight as well. The more older brothers a boy has, the more likely he is to be gay – three times more likely according to some studies. The theory is that mothers develop antibodies against certain proteins produced by boys and that an immune response from the mother actually causes developmental changes in boys that feminize the brain. This brings us to some of the strongest correlative data.

One of the most accurate predictors of a boy becoming homosexual is seen in their behavior as children. Approximately 75% of boys that display childhood gender nonconformity grow up to be gay men. What this means is that if a little boy shows a tendency to consistently display feminine behavior, such as playing with dolls and avoiding rough and tumble play, there is a good chance he will grow up gay regardless of whether his parents force him into more traditional roles or not. It is important to note that not all gay men display this sort of behavior as a child. Again, there are no set rules when it comes to human behavior. The one thing that is consistent is that most homosexual men realized they were gay just before or during puberty and many displayed more feminine behaviors well before that.

More recent data using brain imaging has revealed some significant differences in the way the homosexual and heterosexual male brains react to odors. Two compounds, one found in female urine and the other found in male sweat, are suspected to be pheromones based on animal studies. When gay men smelled the male compound, the part of their brain involved in sexual arousal became more active; the same happened with women, but not heterosexual men. When gay men smelled the female compound, nothing happened, whereas in straight men the brain became active.

Finally, there is significant evidence that nurturing or coaxing of gender roles cannot be done. There are several developmental disorders where boys are born with ambiguous or underdeveloped genitals. For years, castration was seen as the solution, in essence trying to turn them into little girls. The problem was that despite the fact that they were raised as girls, just about every single person turned out to be attracted to women, not men. They had been given a transgender operation without the feminized brain that normally goes along with those who choose to have the procedure. Many of the victims of this misinformed procedure lived tortured lives.

You will notice that I keep bringing up studies on men. This is because the information about lesbians is scarce at best and likely reflects a different determining mechanism than male sexual orientation. I wish there was more I could say about research on female sexual orientation. There are a lot of theories, but not nearly as much data. It does appear as if women set their orientation before puberty, as do boys, and that lesbian women sometimes, but not always, took on more masculine roles as children. But there is little in the way of hard research. The cultural bias is not just against homosexual men.

There is anecdotal evidence that some people who led homosexual lives at some point have switched to living heterosexual lives and vice versa. These isolated cases are extremely rare and do not likely represent *bona fide* choices without underlying biological changes, namely hormone changes. The question of whether homosexuality is something that should be 'cured' ought to be addressed here. The American Psychiatric Association has changed its position on homosexuality several times to reflect its increased understanding of sexual orientation. In the 1950s the APA called homosexuality a 'sociopathic personality disturbance,' which was later changed to 'sexual deviance.' But in 1973 they changed their diagnosis completely, and no longer consider it a mental disorder. Since that time, we have learned a tremendous amount about sexuality. The APA is now a vocal critic of those who claim that homosexuality can be 'cured' or 'reversed' through intense 'reparative' therapy. Clinicians who engage in such practices are certainly on the fringe of the profession and the APA is quite frank about its opinion of their activities, stating that the treatments being performed by these medical professionals are not guided 'by rigorous scientific or psychiatric research, but sometimes by religious and political forces opposed to full civil rights for gay men and lesbians.' They also address the so-called scientific proof that reparative therapy is effective stating:

> This literature not only ignores the impact of social stigma motivating efforts to cure homosexuality; it actively stigmatizes homosexuality as well. 'Reparative' therapy literature also tends to overstate the treatment's accomplishments while neglecting potential risks.

They go on to warn that

> ... psychotherapeutic modalities to convert or 'repair' homosexuality are based on developmental theories whose scientific validity is questionable. Furthermore, anecdotal reports of 'cures' are counterbalanced by anecdotal claims of psychological harm .... Until there is rigorous research available, APA recommends that ethical practitioners refrain from attempts to change individuals' sexual orientation, keeping in mind the medical dictum to first, 'do no harm.'

This is not an activist group by any means: this is the word of the experts. I take them at their word.

Personally, I have always been of the mind that science can provide opportunity for positive change and that we should forge ahead fearlessly into this area of discovery. But that is easy for me to say, since I am not gay and have never been discriminated against because of any particular characteristic I was born with. Nonetheless, I think the fear of scientific advance and the can of worms that could be opened by discovery is something that society should face. People in the gay community know all too well that change can be a painful and prolonged process, and scientists are not unaware and not insensitive to the fact that their work could raise difficult questions. But these are questions that are worth asking if we are to move ahead as a society. The question of whether gay people deserve the same rights and should be treated with the same dignity as everyone else is not a scientific issue, and should not rely on science to be made. That science has been perceived to have intersected with this issue, complicating it for some, is an unfortunate reality of asking challenging questions. But it is the very lack of acceptance of gay people and the prejudice against them that is preventing discovery.

Difficult social issues should not continue to be a barrier to resolving scientific questions about the nature of sexuality. While science that asks the difficult questions about human sexual orientation becomes part of the argument over gay rights, those arguments will have to be won on social fronts, not in the laboratory. There are few if any scientists that would be unhappy if their work illuminated some of the gray areas on social issues, and perhaps that is the best way to view these studies. Conservatives that take issue with such science are more often than not expressing their personal issues with homosexuality, not with discovery.

Even less is known about bisexuality. Kinsey seems to have hit the nail on the head though. He classified sexual orientation on a point scale. In other words, there are shades of gray in between true heterosexuality and homosexuality. There are degrees of bisexuality where most people have thoughts or fantasies of homosexual or heterosexual sex partners, but do not act on them. This seems to be accurate. True bisexuality, having no preference for men or women, but desiring both equally, is extremely rare. More often what you see is people who will experiment with or fantasize about partners outside of their true sexual preference, but in the end settle on one. The theme of not knowing much about the biology of sexual preference is particularly pronounced in this area. We just don't know what would make someone's preference ambiguous, but what we do know indicates that Kinsey was again correct.

If you thought it was rough being gay, imagine what it must be like to be transgender. About half of transgender people report that they have seriously considered suicide at some point in their life. Transgender people often report feeling like they are in the wrong body; that they never felt right even as a young child.

If and when someone comes to grips with the fact that they want to physically switch genders, they have a tremendous uphill battle. It is not just the stigma of being labeled a 'freak', but it involves tough decisions about their future health. For both men and women who are seeking gender reassignment, it involves taking massive doses of estrogen or testosterone and growth hormone. Women who start the therapy to become men experience widening of the jaw, male pattern baldness and hair growth on the body. But with hormone treatment comes serious risk of cardiovascular disease and perhaps some types of cancer. The jaw-dropping moment when it comes to discussing gender reassignment is usually when one broaches the topic of genital surgery. For both male and female surgeries the reconstruction is radical and can lead to long-term complications. So, there is a simple question that is left dangling: To what extent is gender misassignment biological?

It is likely to be biological, since transsexual men and women often talk about feeling that they were the wrong gender very early in life, often well before puberty. In the great majority of cases, there is no evidence of psychological trauma. In fact, most of the abuse these people endure comes some time after they start to display interests common to the opposite gender. Being a transgender person is most often a tragic state to live in. The common self-confusion and ridicule are really quite horrible. But to date there has been exceedingly little research done on transgenders. We know more about diseases that affect a handful of people around the world than we do about transsexuals, and that will remain the way it is because these people are not seen by the general public as 'human' and are certainly not treated with compassion. While research will never make some of the decisions these people have to make in life easier, it could help to ease the blow and explain to them what is actually going on. Until then, the advances made in dealing with transgender individuals will continue to be in the surgical and hormone replacement therapies they go through rather than the underlying cause of their feeling that they were born with the wrong gender.

Until we lift the cultural moratorium on research into sexual orientation and gender, gay people and transgender people will continue to pay taxes and live in our communities without being represented in the research agenda of the nation.

Addicted to love

'Mawidge, the bwessed awwangement, that dweam wiffim a dweam ...' So began the marriage ceremony of Prince Humperdinck and Buttercup in *The Princess Bride*. Most of us know what love is (or at least we know what it feels like), but what makes us love and bond to one person is a mystery. There has been a lot of bile regurgitated over the issue of defining marriage and love, specifically on the issue surrounding whether marriage should be defined as a union between a man and a woman or whether gay couples should be afforded the same rights as heterosexual couples. Besides the pile of civil rights issues that drive the debate, there are some significant biological issues that have been raised about monogamy, and whether there is a biological reason why so many marriages fail. The divorce rate would seem to imply that there is at least the possibility that not all humans are biologically designed to be monogamous for life. Many blame this increasing trend of serial monogamy on immoral values being accepted by society. But it would appear as though a lot of what is being seen is not a case of immoral pursuits or a lack of value on the family unit, but an acceptance that things go wrong in marriages and people grow apart. Is this a reflection of biology?

What most people don't realize (or perhaps want to accept) is that genes do play a part in behavior. It was only about 50 years ago most scientists did not believe that there could be a genetic component to behavior. We now have many examples demonstrating that it is not just environment or experience, but environment and experience affecting genes, brain activity and ultimately behavior. I have already outlined some examples of genes affecting behavior in flies and other animals, but what about the connection between individuals? I would love to be able to report concrete evidence from humans about genes affecting the biology of relationships. Humans suck for looking at how genes affect behavior. There are some examples to discuss, but not many. The real reason is not that human behavior is so intractable to study, it's just that people don't take well to the experiments required to assess such contributions. Removing the brain, slicing it up and staining it for gene expression is particularly frowned upon by most would-be subjects. Also, we have only just completed the Human Genome Project and are still learning about the parts of the genome that affect gene expression and how that affects behavior. So this lack of data is not an indication that these connections do not exist. It just means that we have to rely more heavily on animal experiments and correlative data to discover how this could be happening before we look deeper into human bonding. This is actually the case for most biological problems, but it is particularly true for behavior.

Monogamy is a peculiar behavior in the animal world. The great majority of animals do not practice it, including most primates. However, there are number of notable examples, including some birds, primates and the prairie vole, a rodent that usually mates for life. It is interesting that there are several species of vole in the US, but only one of them is monogamous. Scientists studying the prairie vole found that monogamy, which they refer to as 'pair bonding', is regulated by a combination of very small proteins. One of these small proteins, oxytocin, is routinely given to women in labor because it stimulates contraction of the uterus. In female voles, oxytocin is essential for the formation of monogamous bonds with males. As it turns out, this is also the case for mother–infant bonding in sheep. In males, another protein, called vasopressin, seems to be the culprit. Blocking the activity of oxytocin or vasopressin in female and male voles respectively blocks the ability of these animals to form monogamous bonds with mates. Both of these players need the neurotransmitter dopamine to act, and it is thought that the molecular mechanisms are similar to those that cause addiction. Yes: it is likely that the adage that you can be addicted to love (or more precisely to another person) is correct.

Before we get all mushy let's get gross. Animal models have shown that genital manipulation (How'd ya like to have that job?) results in the release of oxytocin or vasopressin in the brain, followed by a dopamine spike. In humans, sexual activity increases dopamine levels in select areas of the brain. It is thought that this concert of brain activity causes an unconscious learning mechanism where a mate's scent is associated with pleasure. As encounters persist over time, that bond is strengthened. Addiction works in much the same way. Drug use is associated with the extreme pleasure of rapid dopamine release. So, in a sense, one can think of the development of a monogamous relationship as an addiction to the cues that one uses to identify one's mate. This is of course not completely proven, but it's a very likely explanation of why we feel pain when we are separated from our partners for any period of time. It is truly like going through withdrawal.

Humans often speak of missing their mate's smell when they are separated. But humans probably use many cues to bond with their mate, including the sound of their voice and their smile. What about things like the type of cologne or perfume they wear? If a woman wears the same type of perfume a lot of the time, it is completely reasonable that her mate will form a long-term associative memory between that smell and his mate. While there is no good evidence that a perfume or even so-called 'pheromone'-laced perfumes can attract a partner, it is quite possible that people can associate non-biological cues with their monogamous partner.

Monogamy in voles and in humans is most likely an example of convergent evolution, meaning that they evolved separately and reached a similar end. In prairie voles this behavior is controlled by a single gene. In the wild, the sequence of that gene has significant variability, which corresponds to variability in sexual behavior. Some prairie voles are completely monogamous and others fully polygamous, and there is a range in between. When researchers took the gene encoding the receptor for the vasopressin protein from the monogamous vole and put it in voles of another species that is polygamous, their behavior changed. They suddenly started showing preference for a single partner. The difference between the gene for the prairie vole and the polygamous vole was in the area that controls where and how much of the protein is made. So the difference was not in any fundamental function, but in where and how much receptor is made. The same seems to be true for female voles and oxytocin. The two species of vole seem to have similar levels of oxytocin, but oxytocin receptors for the monogamous prairie vole are more widely expressed through the brain. This kind of genetic regulation could also explain the variability in human behavior.

Before we go too far, let's be clear that there is no concrete evidence that the same system that regulates pair bonding in voles is conserved in humans. There is some correlative evidence though, and it is worth considering. Oxytocin levels spike in women with orgasm and vasopressin levels spike in men when they ejaculate. Also, nipple stimulation from breast feeding causes oxytocin levels to raise in women, further supporting a role in pair bonding between mother and child. In a series of experiments where people viewed images of people they were sexually involved with, brain imaging techniques revealed that the same regions of the brain involved in cocaine addiction became more active. These regions have been shown to have a high concentration of oxytocin or vasopressin, or their receptor proteins. And, when men ejaculate, the region of the brain that lights up is the same region that is activated during a heroin rush. It kind of makes you want to try heroin. The point is that pair bonding likely reflects a similar mechanism to addiction and that the cues we use can vary and be enhanced by sexual activity that releases dopamine. In the end, finding a mate and pair bonding is truly 'Ven wuv, twoo wuv, wiw fowwow you fowever,' just like that coke habit.

So now that we know how we form a bond with a mate, all we have to do is choose one. Certainly humans use higher powers than animals in choosing someone to date, right? *Wrong!* It is well understood that men and women treat people of the same sex differently from how they treat people of the opposite sex. This is common to all animals, indicating that in humans this is not a strictly cultural or societal invention. Take mice, for example, who have a specialized structure in the nose to capture fine odors. Mice use these in particular to tell male mice from female mice. If a male mouse encounters another male mouse, they will inevitably fight, but if a male mouse encounters a female mouse, he will try to do her. Simple enough; but remove the ability of a male mouse to detect pheromones and he will try to mate with the other male mouse, even though he can and does use other cues to tell the difference between males and females. For mice, the main cue for mating is therefore scent. Humans are far more complicated, using a combination of cues to tell the difference. But what most people don't realize is that humans use these cues to make judgments about how they feel about people, even to the point where they subconsciously make decisions about mates they are attracted to.

We all like to think that finding Mister or Miss Right has something to do with charisma, charm, personality and looks (we must not forget looks). But to what extent do we really have a choice about who we are attracted to? It is now clear from studies in animals and humans that mating is selective, not just by looks and behavior, but by smell. We've all heard of pheromones that attract animals to one another, and there is some debate as to whether human pheromones exist or still work. What we do know is that there is no universal smell for human men or women that is attractive to the other sex.

We also know that at least in part, we choose mates based on the genetics of our immune system. It sounds weird, but it's true. There is a series of genes collectively known as the Major Histocompatibility Complex (MHC) that are involved in immune response to pathogens. This series of genes happens to have a tremendous amount of variability in it across our species. That means many of the As, Ts, Cs and Gs that make up the genes in this region are variable from one person to another. When it comes to immune response, variability is good because it means we are prepared for more challenges. This variability also plays a significant role in who we choose to mate with. If a person is too similar to you in these genes, you will not find their odor attractive. This is an important barrier to having sex with kin (which somehow disappears in certain southern states). Nonetheless, the same is true if they are too dissimilar to you. This in effect is evolution's way of both preventing you from mating with your family members, but also ensuring that you have enough variability to increase the ability of your offspring to defend themselves from pathogens without straying too far.

Can these odor influences be overcome? Of course. Humans are constantly battling our biological urges and winning. Some of these urges, like attractiveness, are less strong than others, like gender preference, but they do exist nonetheless and create trends in our mating habits.

There is another aspect to this type of work that is also interesting. A study was done on nuns who were presented with T-shirts from men who had worn them for a couple of days. Nice and ripe! When asked to identify the ones that they found pleasant, the nuns generally chose the ones that were from men who were most similar in the MHC genetic region to their father. Before you get any weird ideas about nuns, the question was to identify the one that was most pleasant, not attractive or sexy. This phenomenon likely reflects a need to identify paternal relations and reveals another component of parental pair bonding. As it is far more difficult to identify paternity than maternity, this is an interesting way for evolution to ensure that animals who are related to each other through paternal lines will take care of each other.

Let's get back to who we find attractive. While the smelly aspects of it are interesting, humans are far more visually oriented than mice and dogs. Research has shown that even here there is biology, and it appears to be somewhat different for men and women. For example, studies have shown that men are more attracted to women with a hip to waist ratio of 0.7. This is a weird concept, but it holds up across cultures and time. Even as trends in weight differ over time, the most attractive women generally are the 0.7s. Skinny women, voluptuous women – it doesn't matter, the 0.7 wins. This is likely a reflection of our innate perception of fertility, as prepubescent girls do not form hips and thus do not have 0.7 ratios. The same is true for post-menopausal women who lose their hour glass shape.

While men rely visually on ratios, women tend to rely more heavily on overall body and face symmetry. There are many studies that have shown that women find men with overall symmetry far more attractive than men with subtle deviations from perfect symmetry. This is thought to reflect overall health, but like the 0.7 ratio, it's not a catch-all. So, if you have a droopy eye or aren't as curvy as you would like, don't worry: it's probably just your personality that is keeping you single.

and another thing: genetically modified everything

Go to your kitchen and survey the food in your cabinets and refrigerator. Now roughly estimate how much of the food you have purchased is or contains genetically modified ingredients. If you guessed 100%, you are right. Now consider all of the pets you have had over the years. Dogs? Cats? Canaries? Goldfish? Perhaps a pony? They are all genetically modified. Just about every single food product on the market today and all common pets have been genetically modified by humans.

There are of course some notable exceptions. If you happen to have a refrigerator stockpiled with wild berries, fish and wild game meat, and keep chipmunks and snakes as pets then you're in luck. But no wild corn ever looked like that, bell peppers shouldn't look like bells, and no naturally occurring animal ever looked like the modern chicken or dairy cow. To your credit, most of you realize that Rover wouldn't make it five minutes in the wild. What animal in the wild spends that much time licking its balls?

Human behavior has radically changed the way the world appears, and that starts with the most obvious stuff, like food and pets. We have been modifying food for approximately 12,000 years, so very few people are even remotely aware of how far we have come. It is absolutely baffling to me that humans are so scared of genetically modified organisms, yet we're surrounded by them. When you look closer, the public is not actually scared of all genetic modification of organisms; they are afraid of the use of modern biotechnology, recombinant DNA and transgenic techniques in particular, as it is applied to food and pets. What we need here is clarity about what scientists and companies are doing and why they are doing it, and about what is scary and what is not.

Your pets and just about everything in your refrigerator were produced by selective breeding. This means that organisms that have some quality that people find appealing are bred to enhance that characteristic. So, if you are growing tomato plants and you notice that one of your plants grows really big tomatoes, you breed it with really sweet ones to try to make a line of plants that gives off big, sweet tomatoes. While overly simplistic, this is basically how we have domesticated organisms.

The difference between genetically modified organisms created through breeding programs to selectively distort animals and plants away from what they once were and the use of modern technology is time and control. Most people don't realize that many varieties of plants and animals actually came about through sudden mutation, where someone, usually a farmer, noticed that one particular variety looked different from the rest: yellow and orange peppers for example. This is usually a result of a single genetic change and doesn't seem to bother most folks, but what if humans induced these changes by exposing the plant to chemicals that mutate DNA? Functionally, the plant would do the same thing – just faster.

In some instances, inducing mutations can give a wide assortment of genetic changes across a plant genome and there is little control over the process. Yet, people don't seem to have a problem with this food. It is only when the change is specific that we get uncomfortable. In particular, the insertion of genes into plants and animals, so called transgenic technology, where genes are added to the animal to increase expression of a protein, scares the shit out of people. This kind of pushing and pulling of internal biochemistry is fundamental to science: add a little more or subtract some and see what happens to the rest of the molecules that take part in a biological process.

There are those who would argue that the difference is between what a plant can tolerate through selection and what can be forced upon it. But this argument is really bullshit. For example, take the common bulldog. These animals can't breed normally and their puppies have to be delivered through cesarian section. What's more, few of our modern crops could ever survive long in the wild. In other words, their selection has been anything but natural and the changes that are observed in them are in fact forced through breeding.

While I am not crazy about anyone messing with my food, I also realize that this makes me a hypocrite. Taking a single gene from one species of animal or plant and putting it into another really is well outside of what is even theoretically possible in the natural world. But on the whole the modifications that have been made to food through genetic engineering are quite safe. The real issue here is not with the food itself, or the safety of most of the food, but in the behavior of companies and the US government on this issue, which has given the public the impression that there is something to hide.

I should probably disclose here that I have personally created hundreds of transgenic animal lines. In the course of doing research I created transgenic fruit flies. This is a normal activity in most labs that work with animals. Transgenic plants and animals have become an absolutely essential part of biomedical research, and I guarantee that you will see some Nobel prizes handed out to the people who created the original technologies. Hundreds of transgenic animal lines have been made by removing, replacing or adding genes to animals to model the effects of human disease. Many fundamental insights into biology would not have been made if it were not for transgenic technologies. Some will say that my experience with transgenic technology makes me biased. I couldn't care less. Such views are completely stupid. Having expertise and experience with something does not necessarily render you unable to present truthful arguments about them.

When we talk about bad behavior in this area, we are not talking about academic researchers, who don't usually make GM food. We are talking about large corporations like Monsanto whose activities have been widely criticized, not just because of their behavior toward farmers, but because the purpose of the changes being made often seemed to be directed solely at increasing profit rather than improving the quality of crops. We also have to point the finger at politicians, who bend over backwards to please corporations at the expense of public safety and the livelihood of farmers.

But didn't you just say that the food is quite safe? I did, but that is based upon the scarce information we have on these foods. While there are few cases where one could even imagine the changes being made causing human illness, there is wide concern over the effects that GM plants could have on the environment and rigorous safety tests really haven't been performed.

While close to 30 countries have either banned, limited or required labeling of genetically modified foods, the US has taken a different tack: do nothing. In 1992, the US convened the 'Council on Competitiveness' led by then Vice President

Dan Quayle. At their little powwow, the US government basically decided that genetically modified organisms would not be regulated. This is not to say that they did not put up window dressing. That would be un-American. Technically, the Environmental Protection Agency regulates any crop that can be considered a pesticide. (Yes, there are crops that express toxins that will kill insects that eat them and are thus considered a pesticide.) The US Department of Agriculture regulates testing of plants in fields and the Food and Drug Administration is technically responsible for the safety of GM food.

Some would argue that since there is no evidence that genetically engineered food could pose a health risk, it should not be regulated. But if you don't look, you won't find any issues. More to the point, since there is public concern, having the government regulate genetically engineered foods in a similar way to food additives would certainly have comforted the public to a higher degree and would likely have resulted in the public being more accepting of GM food, further opening the market to smaller biotech companies.

While I am less concerned about food safety right now, I am deeply concerned about environmental impact. Since GM plants can give off pollen and fertilize non-genetically modified plants, there is a risk of contaminating non-GM crops. This is especially true since GM foods usually have no added nutritional value, but are instead designed to resist pesticides and pests, giving them a growth advantage over their non-GM relatives. This has in fact already been seen.

The truly bizarre thing about the discovery of contamination was the way we found out about them: from the seed companies themselves. One pristine example is a Canadian case where Monsanto went on to a farmer's land, took samples and found that he was growing their 'Roundup Ready' canola. Monsanto developed the canola to be resistant to their own herbicide, Roundup. The company then filed suit for patent infringement by the farmer. The only problem was that the farmer had never planted it. The seeds had either blown on to his land or his plants had been cross-pollinated by Monsanto's plants. In fact, through their overzealous patent protection activities against farmers in the US Monsanto effectively established that some of their early products are not easily contained.

Companies also now make seeds for a variety of crops that do not make seeds the next year. This has forced farmers interested in using their product to buy new seed from the company year after year. The USDA and EPA have not required any of these companies to make sure that their products are safe for humans or the environment.

The FDA has only seriously considered data for one GM product, the Flavr Savr tomato, which not coincidentally was the first GM food to hit the US market. This first product was introduced in 1994 and was a total flop. Apparently, in addition to the public being wary of genetically engineered food, the new tomato variety bruised easily and didn't taste very good.

Additional criticism has been ladled on products such as 'Golden Rice,' which was fortified with beta-carotene, a precursor to vitamin A. It was initially hailed as a life-saver for developing countries, where vitamin A deficiency causes malnutrition and blindness. While the company that originally released it claimed that it could prevent 50,000 cases of blindness per year, some back of the envelope calculations by several groups showed that people would have to eat up to 19 pounds of cooked rice per day to reach the recommended daily allowance of vitamin A. Needless to say, such claims did significant damage to the efforts to introduce it to the market and provided more fire power against the corporate entities that have been pushing GM foods. A new variety, 'Golden Rice 2' has up to twenty times higher amounts of the vitamin A precursor. The arguments against such a product fall apart quickly if indeed the new variety is proven to prevent blindness, and there are few scientists that would actually raise questions about the safety of the rice for human consumption. Golden Rice varieties clearly have the potential to be one of the most important GM foods created to date.

Plants are not the only concern of many Americans. There is great fear that genetically engineered pets could cause great damage to the environment if they were to escape. This premise is predicated on the false assumption that genetically engineered animals would have some sort of evolutionary advantage over other animals, or that they would even be able to survive in the environment. This is of course not true as was seen in the first genetically modified pet to make it to market.

In 2004, Texas based biotech company Yorktown Technologies introduced the GloFish to US pet stores. The transgenic zebrafish expresses a gene from coral in its muscles that makes it glow red. Researchers trying to design a fish that would glow when exposed to toxic chemicals developed the GloFish during their research and licensed it. The fish were an immediate success and orders poured in, but some people raised concerns that the fish would make their way into natural populations or had reservations about the very idea of genetically engineered pets. California rejected requests for an exception to the statewide ban on the sale of transgenic fish. One Fish and Game Commissioner actually stated his moral opposition to the sale of genetically modified fish as pets. I guess he's never seen a guppy.

Much of the turmoil could have been avoided if the Federal government had actually taken an interest in the fish. After all, this was going to be the first transgenic pet to hit the market. To the credit of Yorktown Technologies, they approached the FDA, USDA, EPA and the Fish and Wildlife Service looking for them to consider oversight. None of them would even consider the matter. It was particularly strange that the FDA would not consider regulations, because the Acting Director,

Lester Crawford, had stated previously that they would regulate all transgenic fish, whether they were meant to be consumed or not. He made the statement in the context of questions on the first transgenic fish meant for human consumption, a line of salmon that expresses growth hormone at higher than normal levels. This allows farmed salmon to grow faster, thus increasing profits and decreasing the market for wild salmon, even though they are widely regarded as tasting better.

Had any of the relevant agencies even considered the environmental or human impact of the GloFish, they would have found that this tropical fish poses no danger to natural populations because it is not suited to living in US waters, and that the glowing appearance would in all likelihood make it more vulnerable to predation if it were to be introduced into American waters. Non-transgenic zebrafish have been sold as pets in the US for decades and there has yet to be an instance in which the introduction of zebrafish into natural waters has resulted in anything more than predator snacks. Since the red fluorescent protein that is being expressed in these fish has been shown to be completely harmless to many plants and animals, it poses no risk to the environment. Without even considering the safety of these fish, the general public is left in the dark except for the information released by the company.

Thus the lack of regulation by the federal government has helped large corporations like Monsanto bring genetically engineered food to the market while hurting smaller businesses that are producing a harmless product. Luckily, American's zeal for glowing fish outweighed the cries of those who felt that this fish posed an environmental hazard. On the other hand, transgenic salmon would absolutely pose a threat to wild salmon populations if released, because they are endowed with a growth advantage over their non-transgenic brethren.

Many would argue that whether there is regulation or not, the relevant agencies should put out a statement explaining to the public why exactly they do not feel that regulation is necessary. They should explain why they do not see an environmental or health risk in detail and tell the public why the government is not concerned. There is no need for bureaucracy or red tape here: if they are making a decision not to look at something that falls within their jurisdiction, they should say why rather than just wave it on by and remain silent. Certainly this would have worked in the case of the GloFish, and more importantly it would increase public trust in business and in fact increase the number of companies involved, rather than leaving things in the hands of a few giants who have shown hostility towards farmers and anyone who questions the safety of their product.

One area where GM animals could really make a difference is in insect control. The US government has been actively involved in the control of crop pests for decades, most notably fruit flies. To do this they rely on a variety of strategies, but one that few have taken note of is the actual release of insects to control insects. It doesn't seem to make sense on the surface, but it really does work.

Take the Mediterranean fruit fly, medfly, for example. Medfly attack over 300 different kinds of fruit and vegetable. The damage is done when the female fly deposits its egg in the surface of the plant. The US has been aggressively monitoring and controlling medfly for years by growing over 3.2 billion sterile male flies per week in a facility they run in Guatemala. There is a second, smaller facility in Hawaii also involved in insect control. These flies are made sterile by irradiating them. The idea is simple: release sterile male flies who mate with females in the environment and you control the population. In essence, you out-compete the natural population of male flies. This program has been very successful and has kept medfly out of the US for decades, saving billions and billions of dollars to agriculture, which could be crushed by an outbreak. However, this is a very costly process, which could be improved by using GM flies that do not have to be irradiated because they have genes introduced that make them sterile. Scientists involved in research on GM insects have been calling for funding of detailed environmental impact surveys so that the technology can be used. This kind of survey would cost a couple of million dollars, but if shown to be effective, the technology could potentially save the US tens or hundreds of millions of dollars.

The same technology could also be used to develop GM mosquitoes to fight West Nile fever, malaria and dengue fever, saving millions of lives each year. But with no framework in place to regulate these insects and no funding to test their environmental impact, the work will remain stalled. So while the US has been overly liberal inallowing GM crops on the market, technology that would save lives and millions of dollars for the government has gone unappreciated.

Many activist groups have taken a strong stance against the introduction of GM food and pets into the market. They have managed to have strong influence on some local governments in the US and a huge influence in Europe, where plans for the introduction of new genetically engineered products have been halted on several occasions. In the US, however, these groups have been inextricably and erroneously linked to ecoterrorism activities. While I was working at Cold Spring Harbor Lab, one of these groups attacked some of the lab facilities doing experiments on corn, cutting down crops and spray-painting 'No GMOs' on the walls of the facility. None of the crops that they cut down were in fact genetically engineered. They were simple crosses of different mutant corn strains set up by scientists studying plant development.

This act of stupidity gelled many scientists' opinion of these groups. Despite the lab's long and prestigious track record of doing some of the most important plant research, this group failed to understand the distinction between corporate groups making new food products and traditional plant research. The legitimate activist groups would have gained significant legitimacy in the eyes of the scientific community if they had vehemently condemned the activities of fringe groups, but they have failed to do so routinely and loudly. The press has also done a poor job of making the distinction between these groups, making it easier for the US public, scientists and the government to dismiss them all as extremists. To be honest, the government might have done so anyway, as it has always favored the side of food corporations.

What the US needs is a defined system for evaluating the safety of genetically engineered food and pets under one roof. Having it spread out over a handful of agencies is not working and public sentiment has grown against the use of technology. There should be a defined method for testing the safety of food and rigorous evaluation of the potential for harm, if for no other reason than to reduce public fear. This will require some expertise in the government and listening to the counsel of scientists who actually understand what these products are. Such a process would also be good for companies who are developing these products. This is common sense stuff, something that is severely lacking in food regulation.

5

race and other discriminations

"Our tests show you're half white..."

What is race? If you ask five of your friends you will get five dif-ferent answers. Don't worry. You and your friends are not alone. Sitting at a meeting on race and human variation spon-sored by the American Anthropological Association in the fall of 2004, it became clear to me that the experts couldn't define it either, at least as a collective. Granted, if you put a group of social scientists in a room and ask them what color the sky is you won't get a consensus. The point is that race is something we all recognize but have a hard time putting our finger on.

The goal of the meeting was to find some clarity on the issue and to set an agenda for future research and education. While the idea sounds first-rate, it was too lofty and perhaps arro-gant to think that approaching race from a scientific perspec-tive sitting in a conference room at a Holiday Inn could somehow set an agenda or even pose the right questions. As one would predict, the meeting devolved into a competition to see who could present their personal notions using the most words. But the people sitting in that room are not to blame for the chaotic state of our understanding of race. It derives from society's misconceptions and the history of incor-rect notions about the biology of race.

Words like 'nigger,' 'spic' and 'gook', and instruments like the noose and burning cross are what we think of as the tra-ditional tools of racism, but in reality they are the topical uglies, and are hardly the weapons of mass repression that legal and social ideologies pose on people. Discrimination against the vulnerable and exploitation of society's miscon-ceptions about people have been a major theme of human-ity from the start. What is more offensive than outright hate is the hate that co-opts science to justify social order.

False assumptions

One morning, sitting on the 6 train on my way to work in Manhattan, I opened the paper and read the headline 'Alz-heimer's may be Hispanic Epidemic.' The claim seemed auda-cious, and I was surprised that such a dramatic association had not been previously uncovered. After several intense minutes dunking my donut in hazelnut coffee and reading, it became clear that the journalist ... well ... she sucked. Her story was based on a report released by an Alzheimer's disease advocacy group who claimed that Hispanics would be at a greater risk for Alzheimer's in the future because they are going to represent a larger percentage of the population, are increasing in life expectancy, have a lower education level compared to other ethnic groups in the US and have a higher incidence of diabetes. In other words, the Hispanic community is naturally increasing in risk factors for the disease as their proportion of the population and lifespan increase.

I had assumed from the headline that scientists had found a genetic variant that was prevalent in a Hispanic population, but which had not been found in some other population, probably of European descent. This is a fairly common mistake amongst reporters, who interpret the lack of a particular genetic variant to mean that a disease is specific to a particular population and to describe that population by race. There was no reference to new scientific findings in the story, so when I got to the office I went to the actual report and read all eight pages. They read more like a press release than a serious inquiry into the cause of the disease. Indeed, the purpose of the report became clear towards the end, where the advocacy group clearly stated that they were looking to have Congress earmark more money for Alzheimer's research. They had taken a novel angle, but had overstepped the bounds of informing the public about the disease by using fear tactics. The journalist had basically written her story straight from the group's website and press release and given it a nice flashy headline without making any serious inquiry into the report's validity. More importantly, the conclusion that Hispanics will be more susceptible than other populations because they are increasing in proportion to other racial groups and in life expectancy is 100% wrong.

The advocacy group and journalist played an irresponsible duet using the race card to increase interest in a disease and sell a story. The article served no purpose other than to push the group's agenda and scare the shit out of a number of Hispanic readers. Nice work!

One notion that has cropped up among those interested in ensuring racial equality is that there is no biological or indeed genetic basis for race. This is, of course, preposterous. There are undisputed biological differences between different populations in the world – from disease susceptibility to alcohol tolerance – and some of those differences happen to correspond with race. The real question for some time has been to what *extent* are there biological differences between different races. Scientists have naturally been interested in this question and started looking for biological correlates for race some time ago. However, the methods and results of such inquiries have not always been pure in purpose.

What we have learned from the Human Genome Project is that clusters of genetic variants often correlate with the geographical origin of an individual or ancestry. Not surprisingly, these clusters often correlate with individual race. The concept is fairly simple. The DNA of any particular chromosome and thus the genome is made of a string of nucleotides: Adenine, Guanine, Cytosine and Thymidine (A, G, C, T). Some of the positions in the string are variable, meaning that in some people it is a G, while others it could be an A or any other combination. Most of the genome is always the same, regardless of ancestry. Variable sites have a tendency to have a particular letter in a particular population of people. Therefore you can often predict where the person's ancestors were from by looking at the variable regions of their genome. The real surprise has been that when we look at an individual's genome, we often find that their race can also be predicted by ancestry, and that people commonly have ancestry from several populations. So, an African American person with mainly African ancestry will commonly have DNA markers indicating that they also have European, Asian and Hispanic ancestry.

A report that came out in 2005 showed that self-identified race, that is, the race that the person donating their DNA sample tells you they are, correlates almost precisely with geographic ancestry. People who identified themselves as Black were shown to have mostly African ancestry, and the same was true for Asians, Whites and Hispanics. I am white, or so my mother tells me. If you look at my DNA, you will find variations that are consistent with white European ancestry. What we have learned from the human genome sequence is that the borders of distinct populations are pretty fuzzy in most instances. This too makes sense, as populations are not generally geographically isolated and mix pretty frequently. It gives us a continuum of variation across the world with very general lines that can be drawn with little certainty. In other words, you probably couldn't tell that my grandfather was from Belfast, and that my grandmother was from the Ukraine. You might find that I also have Asian and African ancestors as well, but that will not change the fact that most of the DNA markers will show that I am of mostly European ancestry. Humans are spread around the world like peanut butter on bread. Occasionally there is a little dab that sits alone on the crust, but for the most part we cover the surface, with more of us clumping in some areas.

The information we learn about distant ancestry currently doesn't tell us much about biology though. We are a very young species, evolving around 100,000 years ago, and we didn't complete our main migrations until around 3,000 years ago. Since our populations are not generally geographically isolated, we share most of the same variations. The reason we can tell our general distant ancestry is that populations did not always mix. Early in human history, groups migrated out of Africa and settled in discrete locations. The DNA record in our genomes shows us that for the majority of post-migration human history there was little intermingling of populations and that this genetic isolation set the stage for the visible and thus genetic differences between races.

While self-identified race is a good predictor of distant ancestry, it is a crappy predictor of biologically significant genetic differences between individuals. Getting back to our case of bad journalism, identifying someone as Hispanic will not be a good predictor of whether they have a genetic variant that predisposes them to the great majority of diseases. There are approximately 3 billion nucleotides in the human genome, and an astounding 99.9% of them are the same between any two individuals. That still leaves about 3 million letters that will differ. Now, if we look at how those differences vary between human populations in the so-called 'old world' continents, Europe, Asia and Africa, we find that 90% of all human genetic variation occurs on all of them. That means that only 10% of human genetic differences lie between people from different continents. Therefore, there are really very few truly unique genetic differences that correlate with racial identity. That also means that the majority of the differences are common to all races and thus existed before the human peanut butter was spread around the world. This correlates well with the archeological record, which shows that humans originated in Africa and that we lived there for most of our existence.

There are exceptions though, which thus far have been generally associated with disease. This is not because variations that correlate with biology don't exist outside of disease, it is really just an indication of where research efforts have gone. Take sickle cell anemia, the best-known and clearest example of a genetic factor contributing to disease and correlating with one race more often than others. It is caused by mutations in the gene that encodes hemoglobin, the protein that is responsible for red blood cells being able to carry oxygen throughout the body. The mutations cause deformities in the protein and these mutations cause red blood cells to form a sickle-like shape. This causes a host of unpleasant side-effects, including tissue damage, bone deformities and heart disease, because the heart has to pump harder to make up for the lower oxygen capabilities of the red blood cells. The mutation is far more prevalent in populations of African origin than other populations. Some 8% of African Americans carry the trait, and 1 in 500 people are born with the disease. The mutation that causes the disease is thought to have spontaneously occurred several times in history. The reason that this mutation has become so common is that people who carry one copy of the mutated gene are resistant to malaria, which happens to be quite prevalent in the areas where the mutation is prevalent. This is natural selection at work. You give carriers a selective advantage over the non-carriers and bingo: the mutation lives even though it kills approximately 15–20% of the people who carry two copies before they are of reproductive age. This doesn't mean that people of other ancestries can't get sickle cell anemia – in fact, it is found in all races – it just means that they will get it far less often, mostly because the mutation was selected against in populations living in areas where malaria is not a major problem.

Most of the time, there are significant environmental factors that contribute to disease, and it can be difficult to tease these environmental factors away from genetic factors. Here again, race can give us clues. Take for example the incidence of nasopharyngeal cancer. If we take the world population as a whole and do not consider geographic location or race, we find that less than 1 in 100,000 people are diagnosed with the disease. However, if you look at specific populations, you see large differences in incidence. If we use race as a proxy, we see that people of Asian descent get nasopharyngeal cancer more often than people of African or European descent. The incidence of nasopharyngeal cancer is actually much higher throughout Southern China, Hong Kong, Taiwan, Singapore and Malaysia. In particular, if you look in the Guangzhou (formerly Canton) province of China, you see around 80 cases per 100,000 people.

Initially, the incidence was associated with the consumption of dried salted fish, which often contain carcinogens from the drying process. This made sense, since there is also a much higher incidence of nasopharyngeal cancer in the Inuit people of Alaska and Greenland, who also eat a lot of dried salted fish. However, it was still unclear why some people within the same family got it and others didn't, and how some families that consumed a lot of salted fish never seemed to get it. Soon enough, we had an answer. Geneticists examining Asian families who had a history of nasopharyngeal cancer found that there were in fact genetic factors that are more prevalent in people who get nasopharyngeal cancer and that these factors happen to be more common in Asian populations than in populations that are not predisposed to the disease. So, by using race as a proxy for disease risk after a trend was found, scientists were able to show first an environmental factor that contributed to the disease and then a genetic factor. Thus while race alone would be a bad predictor of cancer risk, it was useful in finding the real contributing factors.

Another fine example occurs in Australian Aboriginals. A recent study showed that as a race, they were 62% more likely to die of cancer than other Australian Nationals. If you are of Aboriginal descent, would you be offended if your doctor suggested more frequent cancer screening for you than his patients of other racial origins? Again, race can be a useful proxy in the absence of more specific data.

There are several reasons why any particular genetic variant might be more prevalent in one particular population over another – genetic 'drift,' the appearance of new mutations, 'founder effects' and selection are the four biggies. Genetic drift is a random process where a genetic variant becomes more prevalent in a population. This is especially true for small populations in which a single generation, by chance, inherits a particular nucleotide more than another. If we think about natural history, including human migration and disease, genetic drift likely plays a big part in the presence or absence of a particular variant in some populations over others.

If we go back to our example of bad journalism, it is possible that a genetic variant associated with an increase in risk for Alzheimer's could be more prevalent in one population over another and race might be the best proxy we have for risk of carrying that variant. However, because of the fuzzy lines between populations that proxy will almost never be exclusive for one population. Don't forget: that variant could show up independently in several areas over thousands of years in unrelated events, or it could make its way around the world through population mixing.

Before we get crazy with the details of genetics, let's just take an isolated group of people and show how all four factors can effect genetic variation to better understand the forces at work that created the physical differences between races. Let's say a hundred people who are all neighbors get on some boats and set up shop on an island. Since the group is small, and there had previously been some intermarrying before they left for the island, they will have less variation represented among them than other populations. They cannot possibly represent all of the genetic variants in the human race. Some nucleotides in the genome that are variable across the whole human race will be the same in all or most of them. This is called a founder effect. Now, through genetic drift, some variability of nucleotides could become even less prevalent; after all, it is usually just a flip of the coin that determines whether you get one chromosome or the other from your parents. When you flip a whole bunch of coins at once, some-times you get a lot more heads than tails and vice versa.

Since our group is isolated and because this is one nice island, the people don't generally leave. If a new mutation occurs in a particular individual, it could eventually spread throughout the population over many generations. Now they have a new genetic variant in their midst that is not prevalent through the rest of the world. Let's say that disease or natural disaster kills off a lot of the group. The remaining individuals might have a skewed amount of a particular variant; thus, through no particular biological reason, they might have more of one variant over another. This is a population bottle-neck. But what if somewhere along the line, before the great die-off, a chance mutation that didn't seem to effect the biol-ogy of the people and lurked around in the DNA of the popu-lation actually made those who carried it less susceptible to the disease that wiped out a lot of the population. They pref-erentially survive the disease through dumb genetic luck. This is selection – an extreme form of it, but selection nonetheless. Usually selection takes thousands of years to make any real dif-ference between populations, but this is a small group in an extreme situation. These are precisely the kinds of factors that are at play on a worldwide scale, and isolation of populations, even if not complete and total isolation, will give you skewing of genetics. Those genetic variants often correspond with race, but do not contribute to any appearance or other bio-logical differences between races since most genetic variants do not have obvious functional consequences.

How low can you go?
Remember: race is only a controversial issue because of racism. If it weren't for prejudice, race would be about as interesting as the genetics of hair color or any other run of the mill physical trait. For a time, the dark overtones of racism crept into science, using some legitimate scientific ideas of the time to justify preconceived notions that some

races were superior to others; in particular, that white people were a better 'fit' than people who have a higher amount of melanin in their skin. Without going into a full history of the early abuses of science and discrimination, it is important to understand their context to fully appreciate where we are today in understanding human genetics and the correlations between biology and racial health disparities.

An unhealthy union of science and hate happened in the decades after Darwin published *On the Origin of Species* in 1859. In the US, the Civil War and slavery had just ended, and there were a lot of people who were not ready to accept black people as equals. Justifying their inferiority through science was a natural transition for racists, but it took time because many Christians see evolution as a challenge to the literal word of the Bible, the same book they used to justify slavery (Exodus, Chapter 21, gives a detailed description of the Biblical laws of slavery). The notions of racial superiority and the right to own slaves are cut from the same cloth. The chilling illustrations depicting black people as being more related to monkeys and gorillas than whites, a notion that is still prevalent in some societies, are the perfect example of how people of the day were trying to cope with new scientific ideas and fit them into their own ideologies. Fitting ideology into science and vice versa is always a bad idea, and no better example of it can be found than a movement in the US and Europe called eugenics.

Francis Galton, Darwin's cousin, coined the term 'eugenics' in 1883 as the 'study of the agencies under social control that may improve or impair the racial qualities of future generations, either physically or mentally.' The literal translation of the word from Greek is 'well born.' Simply put, eugenics is the study and practice of selective breeding to improve a population, namely humans. The original idea came from early notions that genetically unfit people were having more children than people who possess desirable genetic characteristics, and so by having the 'fit' reproduce more (positive eugenics), and/or the 'unfit' reproduce less (negative eugenics) you could improve the human race. However, in the early days of the movement, the idea was that environment was causing people to be unfit. This was actually somewhat progressive – it was thought that environmental factors, including smoking and exposure to mercury, would have a negative effect on people and that these effects would be inherited by the next generation. (Over a century after that notion was established, we still find politicians who argue that limiting the availability of cigarettes and exposure to mercury is bad for business interests.)

Many environmental factors can cause mutations that can be inherited. What's more, we now know that pregnant women, when exposed to environmental factors, can give birth to children with a wide array of birth defects without causing mutations. These factors are affecting the expression of genes in the fetus without changing the genes themselves. We still know very little about how the environment affects the expression of genes, so this was a pretty ingenious thought for scientists who had not yet discovered Mendel's work on inheritance. Remember: when scientists came up with this idea, they did not know about genes, DNA being the material of inheritance or about mutations, so they were actually doing pretty well. The problems came when the ideas were actually applied to humans.

At first, it was all about clean living: no smoking, eating well and not masturbating (I didn't say they got it all right). The movement became popular over time, and eventually some in the movement took it a step further, hijacking both evolution and Mendel's rediscovered work on heredity to fuse the new field of human genetics with the eugenics movement. The two were often taught by the same professors at universities, and for many the line between the two was fuzzy or nonexistent. The efforts led to laws allowing sterilization in the US of the 'feeble-minded,' including the mentally disabled, epileptics and criminals, as well as extensive work to document the differences between races, both in physical characteristics and mental capacity.

At least 33 states have had sterilization laws on the books at one point or another over the past hundred years, starting with Indiana's first sterilization law in 1907. Over time, the laws were challenged and even upheld by the Supreme Court in 1927, resulting in an increase in the practice, which continued in the US until the 1970s. Around 60,000 US citizens were legally sterilized between 1907 and 1979. Remember, though, that at the time selective breeding was not considered a completely crazy idea and these sterilization laws were widely seen as a tool to reduce the tax burden of the 'healthy' population by reducing the number of people in correctional institutes and mental health asylums. The movement had many surprising supporters, including Helen Keller and Alexander Graham Bell, although some prominent scientists with a deeper understanding of the complexity of human genetics loudly decried the depravity of the programs. But as we have seen time and time again, governments are not quick to listen to scientists who are telling them what they don't want to hear.

We could dismiss the abuses of the eugenics movement as misguided scientists, applied prejudice or pure hate. We could say that it happened at a different time, when prejudice was more acceptable. The problem is that it is only one of several examples of how government sponsored science and medical programs have preyed on those who are vulnerable, discriminating against them because of their social status. Perhaps one of the most egregious of the examples is the Tuskegee Syphilis Study, which ran from 1932 to 1972. In that program, 399 African American men from Macon County, Alabama, were enrolled in a study to examine the effects of syphilis on the brain. Only they were not told the specifics of the study. They were told that they were in a program to treat 'bad blood,' a generic term used in the poor black culture in the south to describe a host of disorders. The men were not told that they had syphilis, and when penicillin was recognized as an effective treatment for syphilis in 1943 the men enrolled in the study were not treated. Instead, they were left to develop severe brain injury and die of the disease so that US public health service doctors could examine their bodies as the disease progressed. They had effectively become human guinea pigs.

This incident could be dismissed as an isolated incident if it weren't for one simple fact: the Centers for Disease Control, the Public Health Service and many doctors around the country knew about the study. Not only that, when a private physician outside the study saw one of the patients who was suffering from late stage syphilis in 1966 he prescribed a simple battery of antibiotics and was admonished by the Centers for Disease Control for spoiling one of the subjects in the study. Every major newspaper in the US picked up the story the day after it was revealed on the front page of the *Washington Evening Star* in July 1972, sparking widespread public anger. The study was shut down the very next day. In 1997, President Clinton publicly apologized for the study and the mistreatment of those involved:

> What was done cannot be undone, but we can end the silence. We can stop turning our heads away. We can look at you in the eye and finally say on behalf of the American people; what the United States government did was shameful and I am sorry.

To this day, the CDC's website still has inaccurate information about the details of the study.

The thrust of the ill gains from the eugenics movement was achieved by the Nazis after Hitler was elected Chancellor in 1933. What must be understood is that German 'health' initiatives were under way well before Hitler came to power and that eugenics ideas were absorbed into the Nazi party's ideology in the 1920s. German anthropologists were busy categorizing people by genealogy, physical measurements, intelligence tests, race and a host of other methods in line with similar eugenics notions of the day. In other words, what they were doing was not radically different from what eugenics proponents were doing and saying elsewhere. In fact, sterilization laws were enacted in the US over two decades before they were enacted in Nazi Germany. However, by 1933 most geneticists in the US recognized that eugenics reflected an overly simplistic understanding of human heredity and outright prejudice. This, of course, did not stop the increasing popularity of sterilization programs in the US, but they never reached the immense number of victims of the German campaign. The original practices were not held in secret as the infamous death camps were; in fact, Germany proudly presented its eugenics programs around the world at 'health fairs.' Soon after Hitler came to power the first German sterilization law was enacted, which provided for the involuntary sterilization of those 'suffering' from any of nine conditions, including alcoholism, genetic blindness, feeblemindedness and schizophrenia. Hereditary health courts were established to give an air of legitimacy to the efforts. From 1933 until 1945 an estimated 400,000 Germans were sterilized, usually by tubal ligation or vasectomy.

Two years later, the German government passed the 'Blood Protection Law,' which forbade marriage or even sexual relations between Jews and non-Jewish Germans. The Second World War provided the cover that Hitler desired to cleanse Germany of all undesirables. Germany began 'euthanizing' all children born with birth defects in special children's wards at hospitals and clinics. The program was stepped up, authorizing the death of anyone with an incurable disease, starting with the mentally disabled. From 1939–1945 approximately 200,000 Germans were 'euthanized' under state-sponsored programs. With growing public discomfort, the programs became more and more secretive. We are all too familiar with the mass execution of Jews and other ethnic minorities that took place in German occupied Europe through the end of the war.

The impulse to improve the human condition is very different from efforts to create a perfect race of people. Indeed, the principles of eugenics are effectively applied today in the Hasidic and Ashkenazi Jewish communities, which are more genetically at risk for developing Tay–Sachs disease. If you ever saw a child suffering from Tay–Sachs, you would understand the desire to eradicate it. Before a marriage takes place in those communities, the perspective bride and groom have a genetic test for mutations that cause Tay–Sachs. If they both have a mutation in the gene that causes the disease, they don't get married. While this might seem barbaric or crude, it has worked. Tay–Sachs has been virtually eradicated in these communities.

Now, it doesn't always work that way, but you get the idea. A genetic test can be used for people at high risk for a disease to try to prevent the birth of children who will inevitably suffer and die. Those opposed to prenatal genetic testing claim that this is taking a step on to a slippery slope. In a way, they are right. While their reasoning is that it causes people to have abortions, in actuality, genetic testing brings up a host of ethical issues. For example, we now have the ability to select sperm and increase the chance of having a child of a chosen sex by combining the technique with artificial insemination. What if we could select for children with perfect teeth? It would save thousands of dollars on orthodontics, cavity filling and root canals, and children would not have to wear braces, thus lowering healthcare costs for families. The slippery slope discussion is worth having, but in the end we will have all sorts of selection and society will probably shun frivolous application of genetics. That probably won't prevent wealthy idiots from trying it, but you won't see it wholesale any time soon.

Old school eugenics is practiced in one place these days: China. In 1995, a law prohibiting 'dull-witted, idiots or blockheads,' from reproducing went into effect. For some time before that there had been wide application of the 'one child rule,' which left a lot of girls and children with disabilities abandoned or killed. It's hard to tell how widely the new law is being enforced, but human rights advocates have reported that the principles seem to echo many of the ideas put forth by the Nazis. They claim that couples are encouraged to be examined before getting married and are deterred from reproducing if any 'abnormalities' are found. In the case of the mentally disabled, they are being encouraged to be sterilized or to abort any pregnancies that occur. It is completely unclear to what degree forced abortion or sterilization is happening, if at all.

It should be noted that while the modern scientific community has displayed a uniform distaste for the atrocities of Tuskegee, sterilization laws and the entire eugenics movement, US politicians have been slower to come around. In 2001, delegates in the Virginia General Assembly debated issuing a formal apology for the sterilization laws in their state and for the laws that targeted racial minorities. Now, you have to realize that Virginia had laws condoning involuntary sterilization, laws that prohibited interracial marriages, and eugenics laws that targeted Native Americans. Several politicians and a host of citizens commenting on the issue objected to an actual apology on a number of grounds, including the fact that sterilization was legal at the time (I guessed they missed the point of the apology), that the 'racial integrity' laws were not completely misguided, and that recognizing history that Virginians are not proud of will serve no purpose other than to make the people feel bad. In the end, a resolution was passed that fell short of an apology and instead chose to express 'profound regret.' The point is that while scientists have put in place safeguards to prevent the use of new genetic technology for ill-gotten ends, most people are either unaware of the widespread misuse of science to justify hateful policy, or just don't care as long as they can get cheap gas for their SUV.

Disparaging disparities

Unfortunately, the government has taken part in many different abusive research campaigns over the years, including the deliberate exposure of over 16,000 people, including disabled children and soldiers, to radiation and as guinea pigs for vaccines and a host of different chemicals in toxicity studies. For example, children with cerebral palsy were used for radiation experiments at Sonoma State Hospital. There are currently over 1,400 patients buried in unmarked graves at the hospital, many of whom had their brains removed for further study before they were buried. Today, our society has a much clearer understanding of the scope of human rights and informed consent for human experimentation preventing further atrocities ... NOT!

In November 2004, the BBC aired a documentary called 'Guinea Pig Kids,' showing how the New York City Administration of Children's Services (ACS) had been enrolling HIV-positive children in the city's foster care system into experimental drug trials. Children in the foster care system are wards of the state and are predominantly African American and Hispanic from poor economic backgrounds. The drugs, like many antiretroviral medications, are quite toxic, causing severe side-effects. A drug trial on children might seem somewhat cruel, but it is true that children often react differently to medications than adults. However, subsequent reports told of children being taken away from foster parents who had decided that the drugs were not in the children's best interest and of children being force fed the drug through surgically implanted feeding tubes if they refused to take the medication.

While the study design and the drug are part of an NIH sponsored program whose strict guidelines seem to have been adhered to, it brings up an ethical question of whether the government can truly serve as an objective, impartial judge when it comes to giving informed consent in enrolling children in experimental drug trials. Since these children were wards of the state, their foster parents and biological parents did not need to give consent for the children to be enrolled in the trial. It would seem patently obvious to most that foster children are perhaps not the right group of children to use in a drug trial, because of a lack of parental oversight. It also seems obvious that we shouldn't need a law to tell the government that. But here we are again, with yet another case of medical malfeasance where a pinch of common sense could have avoided pain and controversy.

It should be made absolutely clear that incidents like the Tuskegee study and the HIV drug study are rare exceptions in the medical research community. Their impact, however, reverberates for decades after they are uncovered and results in great mistrust in science and doctors, especially within racial minority communities. This is particularly true of the African American community, which is still at the very low end of the health disparities spectrum.

Mistrust is hard to measure, but there are serious indicators that bring the degree of mistrust to light. Take HIV for example. There are widespread fears within the African American community that HIV and AIDS are part of a massive government program to control or wipe out black people. A study conducted by the Rand Corporation and Oregon State University set out to quantify these attitudes within the African American community. They found that more than half of the 500 African Americans that they surveyed, all living in areas with more than 27% African American households, thought that HIV was a man-made virus and that the government is withholding a cure from the poor. Approximately 26% thought that it was made in a government lab and 12% thought it was made and spread by the CIA. Since African Americans make up roughly 13% of the population, but represent over 50% of new HIV infections, with black women making up over 73% of new infections among women, this is not good news.

Racism is still rampant across the world and it doesn't take a genius to realize that African Americans and Hispanics are still treated like second-class citizens throughout much of the US, and that there is still widespread rejection of cultural differences between cultures. Violent racism still persists, as do more passive, less confrontational, forms. Nevertheless, race relations across the world have made significant progress, and science has reflected that progress on the whole. Yet the rich history of racism in the US, combined with high-profile egregious medical projects, has led to cultural divides that manifest themselves in a host of social and medical problems. Mistrust and conspiracy theories are not only limited to the African American community and are not just limited to AIDS. At the same time, it is glib to blame our social injustices and medical health disparities solely on Tuskegee and other misguided medical projects. There are massive health disparities that stem from poverty and poor education within racial minority communities in the US.

To talk about health disparities, it helps to understand the long and short ends of the stick – put more directly, whites versus blacks. According to the Centers for Disease Control, African Americans die of just about every major disease at a higher rate than white Americans. The rate of gonorrhea is 24 times higher in the African American community than in the white community. Syphilis and chlamydia are 9 times higher in African Americans than whites. The disparity is actually increasing for chlamydia, as the rate of chlamydia infections is growing in African Americans at a faster rate than it is for whites, and it will come as little surprise that the highest rates appear in poor inner city neighborhoods. The majority of HIV cases and deaths are African Americans, while whites are below the nation's average. You name the health problem and African Americans are worse off than whites. Three of the top ten causes of death for African Americans are not even in the top ten for whites; homicide, HIV and septicemia (severe bacterial infection of the blood). The only diseases in the top ten for whites that do not appear in the top ten for African Americans are flu, pneumonia and

Alzheimer's disease, all diseases that kill the elderly at higher rates than young people.

These are pretty impressive facts, which raise the obvious questions: are blacks genetically weaker than whites? Are they more susceptible to disease because they did not evolve the capacity to fend off infection as well as whites? Preposterous – but you'd be surprised what people glean from statistics. The fact is, when immunization programs targeted inner cities in the 1970s to lower the rate of measles, the rates lowered for African Americans. More recently, when breast cancer and heart disease awareness programs were enacted in the US, rates lowered as well, though there are still significant disparities between whites and blacks. The point is that a higher percentage of African Americans reside in poor neighborhoods where access to health education, healthcare and healthy lifestyle choices is far more limited than in the burbs, where more whites tend to live. This is a direct result of segregation policies and racism through the last century that are reverberating in society today. Fortunately, the solution to this problem is easy. Create more opportunities for people of poor socioeconomic backgrounds through education programs and safe neighborhood programs and the disparities will dwindle.

There are those who believe that racial profiling in healthcare will only serve to increase the feelings of inequality amongst minorities and add more bad blood between the racial groups, further increasing the mistrust. Agreed; it will. But which is worse for race relationships: health disparities or politically incorrect health policy? Clearly any campaign to resolve health disparities will need a combination of education, outreach and racially tailored policy.

For example, let's consider chlamydia. New infections are found in African Americans at an astoundingly higher rate than any other racial group. What if a comprehensive campaign was started where sex education, condom use, abstinence, parent outreach *and* blanket screening of all young, poor African Americans who come to the doctor was implemented? The program would reduce not only chlamydia, but all other sexually transmitted diseases including AIDS, the eighth biggest killer of African Americans. If a white kid comes to the same doctor's office, don't test them unless there is a medical indicator. This is absolute racial profiling. But it is also the type of program that can empower a community to take their health into their own hands and would pay for itself in reduced health costs once the disparity is reduced. Unlike other outreach and education programs, this one has what lawmakers, community leaders and residents like to see: numbers, in lives saved, reduced health costs and associated drops in unwed mothers and abortion.

Discriminating taste

We now have the complete sequence of the human genome and a good assessment of its variability. This has allowed geneticists to find specific variations in the genome that are associated with an increase or in some cases decrease in the risk of developing specific diseases. They are just starting to uncover the full complement of these disease associations, but it has already proven useful in the development of genetic tests for disease risk. Throughout this chapter, I have talked about disease risk while avoiding telling you what risk really means. In fact, risk is always hard to put your finger on. In some cases, the change in risk is quite minor and therefore not very useful (medically speaking) for telling you if and when you will get a disease. On the other hand, some changes in risk associated with genetic variants represent acute associations with disease: if you have the variant, you will most certainly get the disease. Not surprisingly, most risk is somewhere in between, and is heavily influenced by environmental factors and behavior. This is precisely where the utility of knowing whether you carry a particular variant is useful.

There are now over 1000 genetic tests for diseases on the market, some of which are excellent, while others are nothing more than a fancy way of telling you what you already know from your family history. Let's be perfectly clear: if someone carries a marker for an increased risk of developing a disease, this does not necessarily mean that they will get it. For example, if you have a marker that says you are at greater risk of developing lung cancer, that does not mean that you will get it. That is because these markers indicate genetic risk, but not necessarily cause. In the case of lung cancer, you would likely not get the disease if you don't smoke and are not exposed to smoke on a regular basis. It does mean that if you do smoke, there is a better chance that you will get cancer than someone who does not have the marker, and smokes the same amount as you. The marker does not cause cancer; cigarette smoke does. It just means that your body is more prone to the disease.

All these markers and genetic tests have opened people up to a new type of discrimination based on their DNA. At the time this was written, depending upon where you live, it might be perfectly legal for your boss to fire you or for your insurance premiums to increase because you had a genetic test done. In some cases you don't even have to be shown to have an increased risk. It depends upon where you live. Realize that family history of a disease can serve as an excellent indication of potential risk because you not only share common DNA with your relatives, but often a common environment. So, if your insurance company or employer finds out that someone in your family has a disease, you could be denied promotion or even fired. If your state doesn't have a law against it and your human resources person finds out that you are taking personal time off to care for a sick relative, you could be fired. At the time this book went to press, 33 states had laws that prevent discrimination based on the DNA you were born with, but these are state laws, so some give a little protection while others offer blanket protection.

Since 1996, some congressmen have been trying to pass a genetic non-discrimination law that prevents employers and insurance companies from discriminating based on genetics or from even collecting DNA information for the purpose of determining employment or insurance status. So, if you have a breast cancer test done and it turns out that you are at increased risk, your insurance premiums can't go up. Congress is trying to keep up with science, which has raced right past the laws. This is because of the extraordinary success of scientists. They have moved faster than anyone had really anticipated and the law has not kept up.

The genetic non-discrimination bill passed the US Senate by a unanimous vote in 2004. Progressive politicos had been trying to get the legislation through Congress since 1996, but business interests had held it up. Here was an extraordinary opportunity to pass sweeping civil rights legislation that would protect the rights and privacy of Americans, freeing them to take advantage of the gains of the science they had paid for with their tax dollars. When it was introduced to the US House of Representatives, it had 242 cosponsoring congressmen. President Clinton banned the practice for government employees by executive order in 2000 and President Bush indicated in his weekly radio address that he would sign the bill into law, stating:

> Through a better understanding of the genetic codes, scientists might one day be able to cure and prevent countless diseases .... Employers could be tempted to deny a job based on a person's genetic profile. Insurance companies might use that information to deny an application for coverage, or charge excessive premiums .... Genetic discrimination is unfair to workers and their families. It is unjustified – among other reasons, because it involves little more than medical speculation. A genetic predisposition toward cancer or heart disease does not mean the condition will develop. To deny employment or insurance to a healthy

person based only on a predisposition violates our country's belief in equal treatment and individual merit.

His statement represented a fundamental understanding of the state of genetics research that the President has lacked in several other key areas of science, including stem cells. The bill was never voted on in the House. It sat in the Committee on Energy and Commerce, despite having enough votes to pass and the overwhelming support of politicians, both Republican and Democrat.

Why did the Republican leadership sit on a bill that seemed like such a no-brainer to everyone else? Because business interests got to them. They had been bought. The US Chamber of Commerce and other business groups voiced a host of ridiculous claims about the bill. They said that they were concerned that the bill would cost businesses too much money and cause confusion in trying to adapt to the law. They claimed that it did not provide for a clause that would allow politicians to adjust the protection later when new technology appeared and that genetic discrimination was already covered under existing laws. Finally, they claimed that there was no evidence that genetic discrimination was taking place, and civil rights laws should not be passed unless there is reason to believe that people are being discriminated against. In reality, business lobbyists were coming forward with trumped up concerns that are best summed up as *bullshit*. No Republican representatives openly criticized the lack of action and the bill died at the end of the 108th Congress.

Truth be told, the business lobbyists tipped their hand pretty well by not coming up with reasonable objections. One can only presume that they wanted to be able to discriminate against people. It makes economic sense after all. If you only consider dollars, it would be cheaper for a business to fire a person if they test positive for a risk of developing a disease. It would cost them less to get rid of a person before they develop a disease because then they wouldn't have to pay for increased insurance premiums, deal with future disability payments or find someone to fill their shoes while they are being treated. It will cost them less not to have to be held accountable in court when they are found to be discriminating against employees. Existing laws just don't cover genetic discrimination, and in the same ways that big business preferred to discriminate against the handicapped before the Americans with Disabilities Act was passed into law (which they fought as well), it is appealing for many businesses to get rid of someone who might hurt the bottom line before they get sick. Several studies have already come out showing that people are afraid of genetic discrimination, but would still want to have a genetic test for disease if there was preemptive treatment for it. The great majority of people said that they would be willing to pay for the tests on their own to prevent themselves being discriminated against. One survey found that 32% of women refused a genetic test for breast cancer risk because of fear of discrimination. Further, 83% said they did not want their employers to have access to the results of genetic testing, while 75% did not want their life insurers and 61% did not want their health insurers to have access to test results. The results indicate that an overwhelming majority of the population trust and desire genetic testing for diseases, but do not trust their employers or the insurance industry not to discriminate against them.

In 2005, one of the first pieces of legislation considered by the Senate was the genetic non-discrimination bill. Again, they passed it by a vote of 98–0, sending a message that the rights of the citizens must take precedence over the concerns of business. The House of Representatives again sat on the bill because Republican lawmakers would not bring it to a vote. When this book went to press, the House had not allowed it to come to a vote despite having wide bipartisan support, including that of the President, and we did not hear

Republicans coming to the floor and calling for an up or down vote. Even former House Speaker and devoted conservative Newt Gingrich called for the bill to be voted on, but it fell on deaf ears and the press failed to pick up on his support of it. The bill, along with a number of other important health and science issues, was on the top of the list of legislation that the new Democratic party-led Congress considered.

In October 2005, IBM recognized the value of preventing genetic discrimination and announced a sweeping policy saying that it would not happen at the company. This was the first time any large corporation had come out with such a policy and was an important step forward, but rather than seeing a deluge of companies following suit, they remain a lone giant in the area of new civil rights.

∞∞

There is another twist to the discrimination story that most people never consider. That is – how does research into diseases happen? Or better – how does money get allotted for work on specific diseases? The system seems like it should be pretty simple: money should go to disease research that affects the most people or has the greatest chance for therapeutic payoff. On the whole, this is true, but there is a wrinkle in the system. Money is allocated to government agencies to dole out to scientists. Congress determines how much money agencies get and tells them to keep up the good work, usually. Like most monetary allocation in Washington, there are lobbyists arguing that more money should go the way of their issue. Science is no exception. When it comes to disease research, a good lobbying strategy can mean all the difference. If you get a congressman on your side, you could get a provision put into an appropriations bill that directs the NIH to do research on a particular disease without earmarking it specifically. This seems like a good thing on the surface: make politicians aware of a disease and more research gets done, and maybe some people can get some relief. But when you give more money to research on one disease, you have to take it away from another unless there is an overall increase in funding.

Politicians on the whole are pretty bad at budgeting money. Asking them to understand the nuances of science funding is a tall order. But in their zeal to do good, they could actually be sucking money away from areas where progress is happening at a good pace. Some diseases, like diabetes and heart disease, have a lot of groups lobbying for more money. Money being allocated for this research makes sense because many millions of people have these diseases. What about rare diseases though? Is it fair that clever resourceful lobbying can result in money going to a disease that only 20 or 30 people have?

One good example of this is progeria. You've probably heard of kids with this disease before. They seem to age prematurely, taking on the physical characteristics of octogenarians by the age of 6. The disease is tragic for sure, as the kids who have it usually die of a stroke or heart attack around the age of 13. At any one time there are between 20 and 30 kids with progeria in the US. In 2003, the gene that when mutated causes progeria was discovered and received much fanfare in the press, including speculation about a cure. The truth is, progeria is a fascinating disease, but the odds of a cure coming anytime soon are small at best. Even before the gene was discovered, scientists knew this. Nonetheless, millions of dollars have gone into progeria research. The reason is simple: two doctors had a son who unfortunately had the disease and were very persuasive in getting influential support in Washington. They started an advocacy group and one of them became a fellow at the White House. While the details of how they gained attention for progeria research are unimportant, it *is* important to realize that without these two, the gene would not likely have been found by 2003. Now, one could argue that knowledge of how progeria causes such devastating effects on these children could lend important insight into how normal aging occurs. But while true, one can make similar arguments for just about every disease.

Knowing how things go wrong some of the time can tell you a lot about how things normally happen or how they happen in more common diseases. While it would be hard to describe funding of rare disease research as discrimination, it can definitely be viewed as preferential treatment for those with resources. This happens a lot in Congress, as money is routinely allocated to diseases that directly affect representatives. To be sure, if a congressman had a child with progeria in 1990, the gene would have been found even sooner. It's hard to understand the helplessness of a parent who is trying to cope with a child suffering from a disease that will cut their life short and it is admirable that they try to do everything they can to cure them through research. The manner in which research priorities are set will never be completely fair to many people though.

Another great example is the massive amount of funding that has gone towards researching the so called Gulf War Syndrome. Between 1994 and 2003 the US government spent $316 million researching the disease, which has become a catch-all term for a variety of symptoms that many soldiers from the first Gulf War complained of. However, there is a virtual consensus that the syndrome does not exist. As recently as 2006, the National Academy of Sciences Institute of Medicine concluded that there is no solid evidence that the disease exists. Nonetheless, Congress ended their session in 2006 by calling for an additional $75 million to study it.

On trial

While the academic genetics community is making progress in defining genetic variation across populations, a recent development in the pharmaceutical industry could represent a first step in the wrong direction. In 2005, the FDA approved the drug BiDil for African Americans who have suffered from heart attacks. BiDil is the first drug to receive FDA approval and a US patent for use in a specific racial group. On the surface, this would seem to be a significant medical advance and a first step towards personally tailored medicine. If only it was that simple.

Briefly: BiDil is a combination of two generic drugs, isosorbide dinitrate and hydralazine, that act to dilate blood vessels. This reduces the amount of resistance that the heart has to push blood against. After a heart attack, reducing the load for the heart is obviously a good thing. Dr Jay Cohen originally studied the specific fixed dose combination of these two drugs in the late 1970s. The first trial of BiDil, involving 642 male veterans and showing decreasing post-heart attack mortality compared to a placebo, was published in 1986 in the *New England Journal of Medicine*. However, the effect was said to be of 'borderline statistical significance'. A second article was subsequently published in 1991, comparing the drug to an Angiotensin-Converting Enzyme (ACE) inhibitor, Enalapril. That study concluded that the ACE inhibitor was more effective at decreasing post heart attack mortality than the combination in BiDil. Based on the two trials the FDA rejected Medco Research's application for approval. MedCo did not renew its license on the drug in 1999.

A second company, NitroMed, picked up the license just prior to publication of an article in which the results of the first study were reanalyzed, separating the patients by race. The authors now claimed that the drug showed a greater effectiveness in African Americans than in whites. A new patent was awarded for BiDil in 2002, the first time any drug had received a patent specifying it for a specific race. The patent clearly states that the 'present invention provides for treating and preventing mortality associated with heart failure with an African American Patient.' NitroMed received an FDA 'approvable' letter in 2003, which stipulated that a larger confirmatory study in African Americans was needed. That trial, of approximately 1100 self-identified African American men and women, was already under way, but was halted in the summer of 2004, 6–9 months ahead of schedule, because the drug was shown to be so effective in reducing mortality (43%) and decreased hospitalization (33% reduction of first hospitalization following a heart attack). The results of that study were published in the *New England Journal of Medicine* in November 2004 and attracted tremendous media attention as the first 'race-based' drug.

On 23 December 2004, NitroMed completed its submission of the amendment to its application for approval, and in the summer of 2005 the FDA approved the drug, causing even more fanfare. The scientists should be applauded for their ingenuity and creativeness in reanalyzing their data for the good of a population where there are significant health disparities. After all, it is estimated that 750,000 living African Americans have been diagnosed with heart failure, and that African Americans are twice as likely to suffer from heart failure and twice as likely to die after heart failure than white Americans. In the pharmaceutical industry there is always a rub though, and here we have a shining example of how the rub can raw the flesh.

We have here a patent and the FDA approval of a drug that basically say there is a biological difference between African Americans and other races. Certainly there have been ancestry-specific biological characteristics, including ailments, that have been known for some time. Asians often have lower alcohol tolerance due to lower levels of an enzyme called alcohol dehydrogenase, and sickle cell anemia is found in blacks at a significantly higher rates than in whites. But not all differences are due to genetics. NitroMed is careful to point out that the effects could be due to access to medical care, management of heart failure patients and socioeconomic factors.

Genetic effects will sometimes make a significant contribution to health differences between populations, but don't expect this to be the norm or even common. Other factors, like culture, socioeconomic background and environment, will almost always have a greater effect. While race can be used as a proxy for disease diagnosis, it should be stressed that it is not a very good one because it does not get at the cause of disease. You don't get sickle cell disease because you are black; you get it because you have a mutation in the gene that encodes the hemoglobin protein. The problem is that we don't have that sort of information for most diseases. Sickle cell disease is unusual because it is caused by a single mutation occurring in a single gene. Most diseases are far more complex and often are caused by a combination of environmental and genetic factors or even environmental factors alone. In the case of mortality after a heart attack, it would be great if we knew precisely what to do to decrease it, but we really don't. Diagnosis of the factors that contribute to heart attack and post heart attack mortality would be infinitely more accurate than using race as a proxy for treatment prescription. BiDil might give us a clue to those factors, since we know what the two individual compounds in the drug do. However, there is a bigger issue with BiDil.

First, there has not been a large well-designed study of the effectiveness of the drug in any populations other than African Americans. Thus calling it a race-specific drug might be premature. We already know that disease affects different populations at different rates due to a combination of genetic and environmental factors. African Americans are more likely to suffer congestive heart failure due to hypertension than whites, and whites are more likely to suffer congestive heart failure due to coronary artery disease. Similarly, whites are more likely to suffer from systolic heart failure, which means that the heart is failing to pump blood effectively, whereas African Americans are more likely to suffer from diastolic heart disease, which means the heart is failing to relax after pumping. Thus the effectiveness of the drug could be due to the cause of the heart attack rather than racially delineated genetic differences associated with the drug itself. What if this drug is better suited for hypertension-associated heart attacks rather than coronary heart disease? Further, a race-specific use patent is unprecedented, and has played an interesting role here.

Consider this: results of past drug trials for most drugs have been conducted almost exclusively on white people and have been extrapolated to all races. They are often performed in trials of similar size, so the very notion that this drug will not be useful in white populations is really unfounded at this point. Issuing a patent for this drug in a specific racial group seems premature in the absence of a full comparative study. The company was accused by some of using the racial angle to try to make something out of nothing. But even a cursory look at the data in the African American trial shows a very clear benefit. Here is a plausible scenario though. The original patent on BiDil was due to run out in 2007, so there was certainly an incentive to search for justification for a new patent. By gaining the race-specific patent, the company has limited the use of their drug to one population, in theory. Once word gets out in the medical community that the drug works well in other populations, there is likely to be a lot of off-label use of the drug. Thus, by getting a patent and FDA approval, NitroMed has really gained a much larger audience than it would have you believe. While BiDil may be touted as the first race-based drug, NitroMed has certainly not presented enough evidence that it will only be efficacious in a single population. So, like many scientists, I am torn on BiDil. It is clearly an excellent drug that will save lives, but I find the patent manipulation and the lack of thorough study in other populations for the purpose of increasing profits to be distasteful. Don't worry though. While BiDil is an expensive drug per dose, doctors can prescribe and administer the two generic drugs that make up BiDil at the same concentrations that make up BiDil for far less money.

Unfortunately, the public scientific community and the drug industry are not the only ones interested in race-specific applications of genome information. Since at least the late 1960s the international military establishment has been considering the possibility of biological weapons that are tailored to affect specific racial and ethnic populations – so-called 'Race Bombs.' The idea is simple: if you find a functional genetic difference between two populations, you might be able to create an agent that would have catastrophic consequences for one population and not affect the other. Scientists are currently seeking genetic variants that have functional consequences for biology for the purpose of drug design and disease diagnosis – just one of the goals of the Human Genome Project. These same variants could also be used to design weapons that target people who carry one variant or another.

So, for example, if an army had a biological or chemical weapon that preferentially affected an Arab population, they could have silently delivered it to Iraqi troops and had a distinct advantage over them before a conventional war ever started. (More of an advantage than already existed, that is.) You could justify such an exposure, especially if the agent had no permanent effects (for example, delivery of an agent that gave them diarrhea or made them hallucinate). This could reduce traditional military confrontations and save lives on both sides, but would likely violate the Biological and Toxic Weapons Convention.

There is no public evidence that race bomb type programs are or were under way ... except the words from those in the know. Former US Secretary of Defense William Cohen was quoted in *Jane's Defence Weekly* in 1997 as saying 'I'd seen reports about certain types of pathogens that would be ethnic specific so that they could just eliminate certain ethnic groups and races.' The secretary wasn't just mentioning this in passing either; he was also quoted as saying 'these are the weapons of the future and the future is coming closer and closer.' Now if this was just the musing of a high-ranking US military official one could pass it off as poor judgment in discussing an idea, but that doesn't seem to be the case.

If these weapons are developed, those who would partake in genocide or ethnic cleansing could potentially use them. *Popular Mechanics* quoted Jonathan Moreno, a former official in the Clinton administration as saying 'The South African Defense force conducted research (during apartheid) for the possible development of biological agents that could be used against the black population. They were particularly interested in seeking ways to sterilize women of color.' I take these words with a grain of salt (the military certainly has a history of misleading the public and politicians in the 'interest of national security'), but the fact that they were thinking of these ideas and brought them up publicly is certainly cause for alarm.

There are two topics on race that I have avoided throughout this chapter: athletic and academic ability. Do any of the prejudices hold up? Jimmy (The Greek) Snyder was fired from his sports commentating position in 1988 when, on Martin Luther King's birthday, he said that blacks are better at sports because of slave plantation breeding techniques:

> The black is the better athlete. And he practices to be the better athlete, and he's bred to be the better athlete because this goes way back to the slave period. The slave owner would breed his big black with his big black woman so he could have a big black kid. That's where it all started.

There have been no reliable studies that have demonstrated any correlation between race and ability. Most people now accept that previous studies that have demonstrated that whites and Asians perform better on IQ tests than African Americans are bogus because the tests themselves are fundamentally flawed. Not only is IQ a horrible measure of intelligence, the tests show significant cultural bias and are not fundamentally able to discriminate between racial differences and educational discrepancies. The same is true for athletic ability. Is there a genetic component that corresponds with athletic ability? You bet. But those differences generally have to do with body shape and size and muscle development. Those factors do not correspond with race.

There are of course some exceptions. Isolated populations, like the people who live on Samoa, can skew a group of people toward a particular size. In general, Samoans are genetically programmed to be bigger than the average human. But for larger racial categories there is no biological correlate with ability. There are cultural factors that correlate very well with athletic ability and there is no denying that these cultural factors often go hand in hand with race. Mr Snyder's misconceptions about race are still prevalent today. Sports are often the only road out of poverty and the main activity for black inner city youth, so it is of little surprise that the NBA has a high number of black players. This stuff is obvious to most, but discrimination and discriminatory rhetoric will continue as long as we keep society stupid and allow dogmatic ideology to govern our citizens rather than open discourse and acceptance of cultures. The divisions between racial groups is held in place by poverty and poor education in minority communities. Unfortunately, the trend seems to be headed in the wrong direction, with our political leaders openly lying or distorting science to fit their agenda.

and another thing: global warming

Global warming is a fact. It has been observed, re-created in experimental models, measured and proven. There is now irrefutable evidence that industrialization and rampant use of fossil fuels is wrecking the planet as a place for humans to live. This is not some alarmist hooey; it is a scientific consensus that the surface temperature of the Earth is rising and that we may have already screwed our children's children out of a place to live safely. We are no longer talking about disastrous change in the next thousand or ten thousand years. We are talking about the next 100 years, and we need a radical change in policy to solve the problem (if it is not already too late).

Scientists have been ringing the warning bell for decades now, and the government has been formally warned by the National Academy of Sciences and every other major scientific organization with expertise in this area. Even Hollywood has heard that the average worldwide temperature will rise up to 5 °F by 2100 (although their attempts at creating real stories of warning have resulted in little more than insipid, less-than-thrillers chock full of the latest special effects). That means the virtual disappearance of the ice caps, glaciers receding and the mountain snow pack that supplies much of the world with water supplies shrinking dramatically. But local differences are even more telling. In the summer of 2005, the arctic ice cap shrank to 2 million square miles. That is 500,000 square miles less than the average between 1979 and 2000. This shrinking process is self-perpetuating, since ice reflects light and water absorbs it and heats up. As the water area increases, the heating of the region accelerates.

What does this mean to you? Oh, just environmental and economic disaster. Ocean levels are going to rise, submerging the property and livelihoods of hundreds of millions of people who live in coastal cities worldwide. Within the past 100 years, the average sea level has risen by 4 to 8 inches around the world. Realize that we are at our height of energy consumption and that global warming is accelerating, so most climate scientists predict a rise of 2 to 3 feet in the next 100 years. As Hurricane Katrina made horribly clear, this is going to require trillions of dollars to build sea walls and levees to protect major cities.

Then there will be the huge water shortages. Many cities rely on the annual melting of snow pack for water. In the US, several major cities already consume more water than their local environment can supply. Places like Los Angeles, Las Vegas and Phoenix pipe their water in from hundreds of miles away as their communities recklessly expand and in many cases make little effort to conserve water. In response to Secretary of the Interior Gale Norton's warning that Las Vegas will run low on water by 2025, Mayor Oscar Goodman responded, 'I'm not concerned about the future of Las Vegas as far as water is concerned because I think we'll be able to buy our solution.' But if warming trends continue, that just won't be possible. This is most certainly going to spark more fights over water rights between western states and local municipalities. Of more concern will be the water shortages worldwide, which many believe will cause war between already hostile nations.

Changes in global temperature are also going to dramatically affect the weather. There is already evidence that water surface temperature rises have increased the strength of tropical storms. Several research papers have shown a clear trend of increased strength of hurricanes hitting the US. Hurricane intensity is a measure that combines both the power of a hurricane (i.e. wind speed) and its duration. Researchers have now shown that hurricane intensity has increased on average by as much as 70% in the last 100 years. This is an average, so the appearance of a weak storm today or a strong storm decades ago is not what we are talking about. While these studies are certainly controversial, and the link is now in question by some, the fact that this is even a possibility should scare the daylights out of us.

Fringe scientists and the retarded talking heads of the conservative movement immediately obfuscated the issue by pointing to strong storms earlier in the century and by flatly stating that the science was flawed without actually being able to show significant errors in the analysis. The correlation is quite simple: the National Ocean and Atmospheric Administration showed that the water temperature in the Gulf of Mexico is the highest it has ever been. Hurricanes get their energy from the temperature difference between the water surface and the upper atmosphere. Warmer water fuels storms. So we had more category 4 and 5 hurricanes in 2005 than in any year on record. *Our fault*.

In 2003, the Pentagon released a report entitled 'Thinking the Unthinkable,' detailing the global consequences of not heading off climate change now. They predicted worst-case scenarios in which the US and Europe experience prolonged drought leading to food and water shortages. They talk about devastation to areas prone to ever-intensifying tropical storms. They predict that Europe could see millions of refugees from Africa and other developing regions crippled by famine and disease caused by deteriorating conditions resulting in economic calamity. Finally, they make the grim prediction of war breaking out across the globe, in particular over access to Middle East oil fields, where the US and China square off. In a chilling and frank remark on human nature they warn that, 'Every time there is a choice between starving and raiding, humans raid.'

With the military and scientists understanding and shouting about the consequences of inaction, why do our leaders sit idle? The answer is simple. Conservative groups, funded by large energy corporations, have been allowed to get away with hijacking the political machine in Washington and beyond, sowing doubt about the strength of the evidence for global warming. Basically, if it does not fit their agenda, they bring out a dubiously-funded rent-a-scientist, secure in the knowledge that our supine hacks feel it appropriate to give equal airtime to them and the overwhelming scientific evidence.

What most, including the press, never point out is that the scientific community does not have a rich history of being completely wrong when it comes to making predictions and recommendations about national policy. In fact, there is not one instance in the last 100 years where there has been a scientific consensus coupled with recommendations on national policy where scientists have been wrong. The piss-poor education system in this country has left most people with the inability to weigh the relative strength of precedent and data, and oil and energy corporations know it.

A special shout out at this point should go to Oklahoma Senator James Inhofe, whose continued attacks on global warming science and the environment can be summed up with his statement that research on global warming is 'a gigantic hoax,' which makes him not just wrong, but a disgusting example of leadership gone awry. How can the good people of Oklahoma re-elect this man?

The fight to obscure and deny global warming trends is a logical extension of the efforts that allowed politicians and corporations to protect big tobacco and to attack Environmental Protection Agency reports, the Endangered Species Act and every other attempt to preserve our environment and protect the public. While I really don't believe there is a vast conspiracy, there certainly have been coordinated efforts to undermine science so as to push corporate-friendly agendas. What's more, Democrats and moderate Republicans have absolutely failed to hold their colleagues accountable. The failure of scientists to educate the public, of journalists to report the truth and of responsible politicians to constantly call out and prevent irresponsible legislation from becoming law has left the deathly stench of irresponsible and morally criminal elements in this nation unchecked and rolling in profit.

As energy companies have slowly accepted that global warming is happening, they and their conservative lap dogs have shifted their tactics, now claiming that global warming will be a good thing – warmer weather and higher carbon dioxide will help forests grow while simultaneously giving us more sunny days for outdoor recreation. Seriously. Meanwhile Republicans in Congress continued to prevent any legislation that takes greenhouse gas control seriously despite four of the last five years being among the warmest on record.

In the most recent display of arrogance, President Bush signed an energy bill that did not make a single nod towards developing alternative energy – an effort which, by the way, could create hundreds of thousands of jobs, while simultaneously giving the US a competitive advantage in what is sure to be a rich economic opportunity. Oh, no. His response to $3 a gallon for gasoline after Hurricane Katrina? 'More refineries.' Before signing the bill, the Bush administration delayed the release of an EPA report that showed that average gas mileage for automobiles had actually decreased to levels similar to that seen in the 1980s so that he could sign the energy bill without controversy.

Bush also had the chief of the White House Council on Environmental Quality, Philip Cooney, hand-edit several reports in 2002 and 2003 on climate change to indicate that there is uncertainty on the matter. Mr Cooney, who is a lawyer with an undergraduate degree in economics, specifically changed conclusions stating that global warming was a fact. In 2004, Bush was quoted as saying that we need to study global warming more because of uncertainty; I guess he read the edited version of the reports. Before going to the White House, Mr Cooney was a lobbyist at the American Petroleum Institute. A week after the *New York Times* revealed that he had edited the reports and provided copies of the actual handwritten changes, Mr Cooney resigned from the White House ... and took a job with Exxon. You couldn't make this stuff up.

Over 150 nations have signed and ratified the Kyoto Treaty that aims to reduce greenhouse gases and slow global warming. The only two industrialized nations that did not? The US and Australia. We doomed the entire protocol to failure by not committing our monetary and intellectual muscle to the development of new energy sources. Now, none of the world's main polluters is slated to reach its goals. Part of our responsibility as a superpower is to lead by example. Instead, we drive SUVs. In December 2005, 150 countries agreed to launch formal talks on mandatory reductions of greenhouse gases. The administration refused to enter any 'negotiations leading to new commitments.' Maybe the new Democratic Party-run Congress will do better. It would be hard for them to do worse.

6

bioterrorism and other mayhem

Single pandemic strain looking for unvaccinated host. Prefer older gentlemen who like public bathrooms and intensive care units.

On 14 March 2005, 900 people associated with mail processing centers for the Pentagon started a course of cipro-
floxacin (Cipro), a broad spectrum antibiotic that has been
shown to be effective against anthrax. They were no doubt
in a state of panic even though they were told that the
anthrax detectors are very sensitive and that it was probably
another false alarm. These people are the latest victims of the
2001 anthrax attacks that touched off one of the most chaotic and mismanaged responses to national security in the
history of the US.

False alarms will sound from time to time, and somehow
this has to be considered part of the normal life of people
who process mail. It has been said time and again in the
press that there were only five victims of the anthrax letters.
However, there were actually 23 known infections from the
attacks. The false alarms are a solemn reminder that it could
have been much worse, and that the tens of thousands of
people who have been prescribed medication to prevent
them from getting a disease that they would not have been
exposed to had it not been for one asshole and a poorly
managed biodefense program should be counted as victims
too. More so than any other topic in this book, bioterrorism
is inextricably linked to US politics and misguided policies. In
reality, biological agents have been used many times to
deliberately harm people over the last century, and many
terrorist organizations and nations have possessed bioterror
agents at one time or another, but the real threat of
bioterrorism today does not likely come from these groups: it
comes from within.

There is a laundry list of different chemical, biological and
radiological agents and attack scenarios that have been used
and many more that could theoretically be used to kill or
harm people. Going through the list one by one is daunting
and confusing, even for those who have some understanding of the specific agents and methods of attack. It is impractical and completely unnecessary for the public to wrap their

heads around all the different agents, dispersal methods and outcomes of a bioterror attack – first, because the public education system has left most people ill equipped to understand even the basic differences between agents (viruses vs. bacteria for example), but more importantly, when confronted with such lists and scenarios, it only serves to scare the shit out of them, not prepare them for the realities of this type of terrorism.

I do think that the public should know the basics, particularly the real threat behind different types of attack and what they can do to prepare and react when it does happen. The public should also be aware of what the government should and should not do to prepare for and prevent such attacks, so that they can hold them responsible for getting us ready. If the truth be told, there is little that individuals can do to personally protect themselves from chemical, biological or radiation attacks. Accept that. There is little use in being paranoid about these things. Sure, they are inherently scary and the possibility of future attacks on US soil is far greater than the threat of a nuclear attack from Iraq ever was. But being paranoid or afraid of the future will only get you high blood pressure and lighter in the wallet. It will very rarely protect you, especially from biological weapons, because of the nature of the attack.

B is for Bioweapon, C is for Chemical attack ...

I will touch on chemical and radiological terrorism throughout this chapter because they are erroneously linked in the press as being related to bioweapons. In reality, most chemical and radiological weapons are more similar to conventional weapons in their effect than either are to biological weapons. The first and most noticeable difference is in the method. The perfect biological terror attack would be silent; the perpetrators could be on the other side of the world before the first signs are noticed, whereas most radiological and chemical attacks are noticed almost immediately. Unlike the traditional responders to terror – fireman, police, HAZMAT specialists etc. – who are responding to a known event, the first responders to bioterror could be those charged with recognizing that something has even happened. This means that the first line of defense is doctors, nurses and other health professionals within our public health infrastructure recognizing unusual symptoms or an odd number of people with similar symptoms coming in to see them. The anthrax letters had a note pointing out that the powder was dangerous, but an attack aimed at mass casualties would not likely have such a warning. The initial symptoms for most agents that would be used in a bioterror attack are not very different from the symptoms of common diseases – in many cases the common cold that health professionals see every day. So they would need to notice a trend or significant spike in people coming in with specific symptoms. Not only that, it relies on communication between hospitals to see if there is a trend emerging. If the agent is communicable, then you will get waves of people: those who were first infected from primary exposure to the agent, followed by those that contracted it by coming into contact with others who were initially exposed. The reality is that an attack would not likely be recognized until after it is already too late for the first victims.

Unlike traditional terrorism, the preparedness of primary responders and public health infrastructure has a direct effect on the number of people who are infected and die from bioterror or any highly infectious disease for that matter. If hospitals are alert, prepared and coordinated, it will likely make all the difference in the world, whereas there is little anyone can do if a maniac decides to blow up a building or release a chemical into an environment. It is the ongoing neglect of the public health system that is our biggest threat in handling bioterror. Our public health infrastructure has been left to fall into disrepair for so long and the response to direct threats has been so slow that several years after the original anthrax letters and over 10 years since the attacks on the Tokyo subway system it is scarcely better equipped to deal with a large-scale bioterror attack.

Ninety-five per cent of US hospitals are privately owned, so corporate interests compete with quality of care and public health preparedness. There is little or no incentive for a corporate-run hospital to go the extra mile in preparing for large disasters. Bioterror preparedness does not help the bottom line, so these corporate entities are not going to do any more than the minimum that is required of them by law, and even then, with one third of hospitals running in the red, they might not be able to afford to go the extra mile. This is particularly true when it comes to quarantine beds. There are no hospitals in the US that could take in and quarantine hundreds or thousands of exposed individuals in the event of an attack with a highly communicable agent. Public health officials refer to this as surge capacity, and our government is well aware of the situation.

To truly understand biological, chemical and radiological terrorism, you have to try to think like a terrorist. This is an absolutely disturbing exercise, but a useful one for preparing countermeasures. First, one has to understand the goal of terrorism. Most people erroneously believe that terrorism is designed to kill lots of people. This is not supported by most terrorist actions, but is what we spend most of our effort preparing for. It is in fact smaller events, like small, improvised bombs that make up most terrorist actions. The true impact of all terrorist actions is psychological, and this is precisely what terrorists are aiming for. Terrorism is a very effective form of theater that plays on people's fear of uncertainty and vulnerability. It causes a lot of social upheaval and economic damage to entire regions and sends governments into chaotic frenzies that usually do nothing more than further scare the public and restrict freedom of the citizenry. If in the process they make it somewhat harder to pull off an attack, that is a much needed bonus, but for all of the posturing and regulations and ridiculous bureaucracy that have been put in

place since September 11th 2001, someone could still drive
into any major city with a truck bomb like Timothy McVeigh
did and would not be stopped.

It is important to think rationally about chemical and biological terrorism, without the rhetoric that we hear from
Washington. Time and time again we hear Bush administration officials and congressmen talk about the threat of these
agents, but almost never display any real understanding of
how these things really work. So let's think like terrorists.

The general public has been bombarded with misinformation about potential radiological attacks in the form of a so-
called 'dirty bomb.' Most scenarios played out by the press and
politicians erroneously state that people would die of radiation
sickness. Not true. Uranium and plutonium do not put out a lot
of radiation when they are just sitting around. The confusion
comes from people not understanding that you need to have
nuclear fission to cause the type of damage we associate with
these heavy metals. Also, the government has done a horrible
job of telling people that they *can* survive a nuclear attack. The
most important thing to know is that if you see a mushroom
cloud you can survive, because you were not likely hurt in the
initial explosion. The danger at that point is the radioactive
material in the cloud. Next, you watch to see which direction
the plume is moving: if it is coming towards you, walk perpen-
dicular to it. If you do this, you can easily walk out of the area
that is going to be exposed to radioactive material.

With dirty bombs, a person would die of heavy metal poisoning, not radiation poisoning, because our bodies are not
designed to clear heavy metals, like uranium and plutonium.
This is pretty much the same as mercury poisoning. The big
effects of such a bomb would be the environmental, psychological and economic consequences associated with having
to clean this stuff up and from people panicking from the significant explosion needed to spread radioactive material.

When the Center for Strategic and International Studies
performed a simulation of how the public would react after a

dirty bomb was detonated in downtown Washington DC, they found that employees would not return to work or even enter the city because they would not perceive it as being safe. The public would essentially not come to the area, even after the radioactive material was cleared. If a significant dirty bomb were detonated in a large American city the economic cost to the country could be hundreds of billions of dollars, making the anthrax letters look like loose change. To what level would you have to clean the building to assure the public that it is safe to return? It is certainly lower than the standards that the government proposes for safety, and some watchdog groups claim that the relaxed standards are designed to protect the powerful nuclear industry, which some claim has the Department of Energy in its pocket.

Hospitals are now training staff to deal with dirty bomb scenarios, but there is no way to prepare for widespread *War of the Worlds*-style panic. It is surely easy to make such a device, but getting radioactive material is much harder. There are reports that say that there are vast quantities of nuclear material that is available on the black market from the former Soviet Union and that the stockpiles that are currently stored in eastern bloc countries are poorly protected at best. But one would presume that it would still be pretty hard for the average maniac to get a hold of these. Why go through the trouble of stealing radioactive material just to poison people and make a mess of the environment? You could just open up a chemical plant and slowly pollute communities and waterways while making profit. In other words, dirty bombs are scarcely better than regular bombs for killing people. In the end, a dirty bomb is an inefficient scenario for terrorists to use if they intend to kill people. They are better off just using a regular bomb. Anyone with the will to pull off a terrorist action knows this as well.

The same is true of most chemical attack scenarios, in the form of sarin or some other agent. The most severe effects of such an attack will likely be economic and psychological, as it would be exceedingly difficult for a terrorist to acquire, store, transport and disperse enough of these agents to cause mass casualties (in the hundreds or thousands). There are a few scenarios where release of a chemical weapon in a coordinated fashion could cause hundreds of deaths, such as in a subway system during rush hour, but it is pretty hard to do. If we accept that an essential part of terrorism is dramatic disruption and fear, then we quickly realize that most dramatic chemical attack scenarios are no more effective than bombs.

Bioterror, on the other hand, has the advantages of being difficult to detect, relatively easy to acquire (compared to plutonium), simple to transport, possibly communicable and useable in any number of ways. There are a wide variety of methods of using biological agents to harm people, including indirect exposure (as we saw with the anthrax letters), attacks on the processed food supply, agricultural bioterorrism, and of course direct exposure, like spraying an agent on people or injecting them with it. This is the scary stuff, and the public should be very concerned that we are not ready for a large-scale attack, or any public health emergency for that matter.

Because a bioterror attack can take place in so many possible forms, we should get to know the basic types of agent that could be used and then focus on the most likely scenarios. The big bioterror attack that we all fear takes advantage of the innate properties of organisms to cause damage and uses lessons from natural disease outbreaks to maximize its impact. Natural infectious disease outbreaks have caused massive die-offs in the world on several occasions through history. However, these have never been caused by people designing or utilizing an agent specifically for the purpose of causing death, so it is still unclear as to whether such large attacks would succeed. (That is unless you count the deliberate infection of American Indians with smallpox by early American settlers through the gift of infected blankets.) On the contrary, most of these outbreaks have been caused by the ever-expanding human population and the lack of sanitary conditions through history. As humans overcrowd or encroach into new regions, outbreaks occur. Poverty and unsanitary conditions, which usually come as a package deal, create an environment where viruses and bacteria can thrive. So bioterrorism has an uphill battle, because conditions are not optimized in most cities and communities for this type of spread. Thus it is important for a terrorist to maximize the effect of an agent. This is particularly difficult to pull off without extensive knowledge of how to handle agents and usually requires quite a bit of training.

Choices, choices

Discussing the complete list of possible agents that can be used would be silly, but some good examples will give you the gist. I don't want to provide a cookbook for terrorism, but discussing types of agent is a useful way for us to start to understand it.

The agent chosen by a potential bioterrorist depends largely on his intent, skill level and access. With the proper motivation, no agents are really impossible to acquire. Some agents, like smallpox, are really hard to get, but many agents can be acquired quite easily.

Weaponizing and spreading an agent to an individual or population is another matter. It can require great skill, which only a limited number of people have. This is a key point: the use of individual agents is tightly linked to and limited by the potential methods that can be used to spread them and the way they must be treated before they can be spread. We should probably start off with the prima donna of all bioterror agents, the bacterium.

Easy to grow and manipulate, bacteria can really do a number on people and the food supply. Our old friend anthrax is the BMOC of bacteria in the bioterror world. At least four groups have tried to use anthrax to kill people, but only one has been successful: the letter sender from 2001. The bacterium itself has a life cycle that includes a hardy spore stage that, if weaponized, can be quite deadly. Unweaponized spores on their own are highly unlikely to kill many people if dispersed, but are still quite dangerous. Other bacteria, like the species that cause plague, typhoid fever, food poisoning and dysentery, are also good candidates.

Next are the viruses. While not technically organisms, they can be quite nasty as biological weapons because they can be easily communicable. Viruses depend upon the cells they infect to help reproduce. Outside of a cell, a virus is quite inert and has a protein coat that protects it from the environment. When viruses come into contact with a host's cells they inject their genetic material (either DNA or RNA) and then take over the cells' molecular machinery to make more of the virus.

Several viruses that cause disease are easy to produce in the lab in large quantities. However, they are often more delicate than bacteria, and thus can be harder to weaponize. People have been deliberately infected with HIV in a few biocrimes, and others like Marburg virus (hemmorragic fever), Ebola, influenza and of course smallpox are potential bioterror agents. Also, unlike bacterial diseases, there are few cures for many viral diseases. They are nasty with a capital N.

The next category is toxins. These are poisonous chemicals produced by organisms. To extract these, you have to have some chemistry skills and really want to use this type of weapon. They are very difficult to purify and are not generally toxic unless ingested or injected. They are never communicable. The major player is ricin, which is easily isolated from castor beans and has been used a number of times. Then there is botulinum toxin – that's right, Botox. Unlike the stuff injected into the expressionless dilatants that saunter along Rodeo Drive, the stuff we're talking about is far more concentrated and very deadly.

Parasitic worms and insects can be used as agents and to disperse agents. There is actually a case where someone was deliberately infected with a parasitic worm that normally only infects pigs, but this is an impractical and unlikely scenario for a larger population. The growth and release of insects could potentially cause significant economic damage to agriculture, but is also completely impractical. Some have speculated that the introduction of a new species to an area that will displace crops or other wildlife could be used. But it lacks the drama required for effective terrorism.

But insects can be used as an effective disease vector and have been used to deliberately spread plague in the past. There are two types of plague – bubonic and pneumonic – both of which are caused by the bacterium *Yersinia pestis*. Bubonic plague is spread by fleas who bite infected animals and then humans, and pneumonic plague is spread by breathing in aerosolized bacteria. Both are easily cured by antibiotics, but pneumonic plague is the big fear because it is quite contagious. The fear of infected biting insects like fleas and mosquitoes being used as bioterror vectors comes from the fact that the Japanese actually used plague as a bioweapon in their invasion of China prior to the start of the Second World War by spreading infected rice for rats to feed on and releasing infected fleas. The tactic worked, and many Chinese soldiers and civilians died of plague, but the unpredictability of using an insect vector makes it impractical, as the Japanese found. Many of their own troops became infected as well.

Fortunately, modern living conditions are also far different than they were back when bubonic plague was a big problem. The bacterium is easy to find, however: as one scientist pointed out, if you culture bacteria or collect fleas from prairie dog holes in the US, you'll find it soon enough. Most people don't realize that there are still plague cases in the US every year. Worldwide, there are about 3,000 cases a year, but this number varies widely. The combined problem of growing large quantities of fleas, inappropriate conditions for flea survival in the US and the fact that it is quite curable makes this an unlikely method for use as a bioterror agent.

Mosquitoes have been proposed as vectors as well, since they can carry West Nile virus, yellow fever and dengue. The US actually built facilities to grow large numbers of mosquitoes for this purpose before it signed the Biological and Toxic Weapons Convention in 1972, but again a terrorist group would be stupid to go in this direction.

Fungus and other plant pests have been proposed as good candidates for agricultural bioterrorism. Again, growing enough and disseminating them so that it infects enough corn, for example, to have any kind of substantial impact on an economy – even a regional economy – would be unlikely if not impossible for a terrorist organization to pull off and won't have the immediate terrorizing effect of a bomb or nasty chemical release. The point is that just having an agent doesn't mean that you will be able to cause terrorism unless you know how to prepare and deliver it effectively. Insects and plant pathogens are clearly not the best way to cause theatrical fear.

A biological agent is not necessarily dangerous unless it exists in a form that eases dispersal. So terrorists not only have to understand how to handle the agent, grow it, concentrate it and store it, they have to understand how it will act outside of the lab dish, and how and where it can be spread. If they choose an agent that is not communicable (anthrax for example), then they have to spread it to more people. Not all agents are contagious, and even if they are, they can be quite safe to work with in a lab; that is until they are weaponized, at which point they become highly dangerous to manipulate. But even contagious pathogens have to be dispersed through at least one infected person or animal. Those initially infected with a communicable agent can in a sense become the main form of dispersal.

In general, the most feared method of bioterrorism is aerosol dispersal because it is a very good way to get a lot of people sick. This is how five people died from the anthrax letters. The pure form of the powder mailed to the Senate offices in Washington DC contained spores that were small enough to escape through the microscopic holes in the paper and envelopes that contained them. Spores and clumps of spores in this size range of 1–5 microns will float through the air for some time and can be breathed in. There are one thousand microns in a millimeter, so you need an electron microscope to even see spores this size. Since anthrax is not contagious, only people who breathed in the spores got an inhalation infection. We can consider the inhalation anthrax infections to be aerosolization, even though it was not likely the intentional effect.

Another way to spread bioterror agents is through direct exposure. The most common way to infect someone directly with an agent is injection. There are several cases of people being injected with HIV infected blood or *Salmonella typhi*, the bacteria that causes typhoid fever. This method can't be used for mass casualties because it is very unlikely that people would be unaware of the fact that someone had jabbed them with a needle. And while a sociopath could run around sticking people with a needle, that scenario is best left for Hollywood. With some toxins and bacteria skin application is a possibility, but even though it gets past the challenges of aerosolizing an agent, it is not likely to cause mass casualties unless it is a highly contagious pathogen.

Indirect exposure is a real possibility, though it involves relying upon a person to pick up the agent through their own actions, whether by eating food, touching an inoculated surface. The problem with indirect exposure is that most agents don't survive well outside of optimal conditions. But where there is a will, there is a way. Consider this scenario in which someone places an agent on a heavily used doorknob. In most cases, this would fail to cause any illnesses. But if enough doorknobs have the agent applied, it could work. What a person is relying on is that people are kind of gross. We have not learned general lessons on cleanliness and are constantly putting our hands near our face after touching doorknobs, hand rails, and computer keyboards.

We eat without washing our hands first and every one of us has noticed someone not washing their hands after using the bathroom. The indirect scenario could capitalize on the same poor hygiene that makes us catch illnesses now.

One form of indirect exposure involves inoculating food. This method has actually been used successfully to make a large number of people ill. Both parasites and bacteria have been deliberately applied to food to make people sick.

In 2005, outgoing Secretary of Health and Human Services, Tommy Thompson, expressed his surprise that the US food supply had not been attacked. He caught a lot of flak for his candor, and was accused by some of being spiteful because he had not been offered the Secretary of Homeland Security job that was being vacated by Tom Ridge. (Remember the mass exodus of cabinet members from the Bush administration in early 2005.) The press criticized him for pointing out the vulnerability of our food supply and possibly giving terrorists an idea. This is of course stupid. If terrorists are relying on administration officials for ideas, then they are likely to be in more danger of hurting themselves with agents than having any real effect on the food supply. Within the scientific community and the NIH, Mr Thompson was considered an ineffective leader, but to be honest, his admission was refreshing. Our food supply has been increasingly centralized over the past decades, and we import more food into the US than we ever have. US food inspectors are so over-stretched that very little food is effectively monitored or even inspected. But let's be honest with ourselves here: even if there were thousands more inspectors and security was truly efficient, there would be no way to police the food supply on any appreciable level. It is just not possible to easily detect bioterror agents on or in food or drink products, so we will remain vulnerable in this area.

Since real experiments demonstrating vulnerability are impractical, scientists often turn to mathematical modeling to try to calculate vulnerability. In 2005, a paper was published in the *Proceedings of the National Academy of Sciences* that pointed out the vulnerability of our milk supply to terrorists interested in using botulinum toxin. Their mathematical modeling of an attack on the milk distribution chain led them to conclude that a couple of grams of toxin sprinkled into a vat of milk at a distribution center could cause hundreds of thousands of deaths – mostly of children, since they consume more milk and are smaller than adults. While the analysis made some presumptions that were overly simplistic, it brought up an important point: that the centralization of our food distribution makes us more vulnerable to large-scale attacks. The bright side is that it might also make it easier to guard the food supply. Since the US consumes approximately 20 billion gallons of milk each year, the model was appropriate to demonstrate how we are vulnerable. But the scientists were heavily criticized by officials at the Department of Health and Human Services for supplying a roadmap to terrorists. This is absolutely untrue. If scientists don't point out our weaknesses publicly, there will not be any action taken by the government to protect citizens. This is especially true since government has actively ignored the recommendations of the scientific community on many security and public health issues.

Some have mentioned the introduction of biological and chemical agents to the water supply as a means of mass infection. The risk of this type of event has been deemed highly unlikely because chlorination, filtration and other chemical treatments to all metropolitan water supplies would kill off almost all agents and the dilution factor is similar to the principles that allow industrial companies to dump toxins into the supply ... 'dilution is the solution to pollution.' 'Experts' have claimed that it would simply be impractical to introduce enough stuff to make this means of dispersal effective. *Not true!*

While the dilution factor for some large municipalities would make it difficult to attack a city (New York for example), most water supply systems are much smaller and thus more vulnerable. Recently, government-funded scientists ran the numbers. They found that the presumption was wrong and that the water supplies that feed the majority of the US are quite vulnerable to a variety of toxins, bacteria, viruses and chemicals. First, water treatment facilities do not test for most agents that would be used in an attack, but even if they did, it is highly unlikely that they would catch an attack because they are not tested constantly, just routinely. An agent introduced into the system would be completely flushed out of most municipal systems in a matter of a few hours, and even if they did detect it, there is no reasonable way to stop people from using their water fast enough to save them short of shutting down the entire water supply, which many cities actually don't have the ability to do quickly. Also, the chlorine that is added to the water to kill off bacteria and viruses can be easily inactivated by adding simple chemicals that can be easily acquired.

When they looked into the amount of agent that would have to be added to kill or make thousands of people ill, they discovered that for some agents it wasn't truck loads, but a couple of pounds. The scientists claimed publicly at a meeting that some of the agents they looked at can be easily purchased in large enough quantities to pull off an attack. Finally, they concluded that since the water supply is considered safe by the government and health professionals, no one would ever suspect that this is what had caused the event and by the time they figured it out, thousands could die or be injured and all of the actual chemical would be washed out of the system. Don't bother panicking and only drinking bottled water though. The plants that bottle water are not much safer, and are in reality probably far more vulnerable.

I realize that some will think that I implied that large municipal water supplies were safe from terrorism ... also *not true*. For example, the New York City water supply is fed by two very large old pipes. If either one of these pipes were significantly damaged in a conventional terrorist attack, it would destroy the city because there would be no way to compensate and get enough water flowing in through the other remaining pipe. In essence, New York City would be shut down and there would be no short-term way to fix the problem. This would in effect be the largest economic disaster in the history of the world, and while not bioterror *per se*, the effect will essentially cause a biological disaster by restricting the water supply. The city has been building a third pipe for some time now and has recently stepped up the building process because of the public revelation of the city's vulnerability, but the process of building these giant water supply systems is difficult and long. Completion of pipeline number 3 sometime in the next decade will be one of the largest municipal projects in history.

Before we consider the anthrax letters, we should look at a few other cases where terrorist groups have tried to use biological agents to get a sense of what we might expect in the future. In 1994 and 1995, the Japanese cult Aum Shinrikyo dispersed sarin nerve gas in a series of chemical attacks including a coordinated release in the Tokyo subway system. Altogether, 12 people were killed in the subway attacks and close to 4,000 people were injured. Also, seven people had been killed in a previous sarin attack by the cult in 1994. In addition to chemical attacks, it has also been reported that they unsuccessfully tried to use anthrax and botulinum. Later it was revealed that the anthrax that they had obtained was actually a non-pathogenic strain being used to develop vaccines. It should be noted that they also tried to acquire Ebola virus and the bacterium that causes Q fever and cholera, but were either unsuccessful or were unable to use it.

To succeed in their attacks, it has been reported that the cult had to recruit many scientists, most of whom were considered failures in the scientific community. Despite having some expertise, the challenges in making biological weapons were too great for them to overcome.

The successful chemical attacks created significant changes in Japanese society and served as a warning that fringe groups need to be kept under control. Most people don't realize that the cult had construction contracts and access to explosives. Pound for pound they would have killed a lot more people if they had made bombs rather than sarin gas, illustrating the inherent difficulty of using these types of weapons.

Another group, called Dark Harvest, protested the British government's failure to clean up Guinard Island, where it had tested anthrax during the Second World War, by taking soil from the island contaminated with anthrax and dumping it in front of the British biological agent research labs at Porton Down in 1981. No one was infected by the soil.

The only successful bioterrorism attack on US soil before the anthrax letters was carried out by an Oregon cult called the Rajneeshees in 1984. They deliberately contaminated salad bars with *Salmonella typhimurium*, which resulted in approximately 750 people getting sick with salmonella poisoning. There were no deaths, but 45 people did have to be hospitalized from the attack.

Outside of the activities of these groups, several groups have shown interest in acquiring bioterror agents, including a few bumbling white supremacists and assorted other miscreants. There have been some cases where individuals have been deliberately infected with agents, but these so-called biocrimes aren't really terrorism because they were aimed at an individual and there did not seem to be any political, economic or societal motives.

So where are the terrorists who are plotting to use biological weapons against people? Well, it appears as though there are, or were, a few of them before the anthrax letters and those that used them had significant difficulty or completely bungled their efforts. Which brings us to today.

There is no question that there has been a serious increase in anti-American sentiment since the start of the second war in Iraq. The administration has clearly misled the public on a number of issues leading up to that war, including grossly exaggerating about the biological and chemical weapons capability of Iraq. A former aide to Secretary of State Colin Powell called the weapons of mass destruction speech delivered to the United Nations the lowest point in his life. There are many who believe that the unconscionable behavior of the Bush administration has created a new breed of terrorists who are intent upon using any means possible to hurt Americans. More people died in 2005 from cigarette smoking, accidental gunshot, slipping on ice, dehydration, overhydration, methamphetamine abuse, suicide, HIV or the lack of response to Hurricane Katrina than biological weapons, yet the US spends billions of dollars each year preparing for it. Finally, more Americans have lost their lives in the Iraq war than died on September 11, 2001.

2001: the year we lost contact

YOU CAN NOT STOP US. WE HAVE THIS ANTHRAX.
YOU DIE NOW. ARE YOU AFRAID? DEATH TO
AMERICA. DEATH TO ISRAEL. ALLAH IS GREAT.

So read the letter that was opened on 16 October 2001 in the office of Senator Tom Daschle. The letter was decidedly more aggressive than the letters mailed to the *New York Post* and then NBC anchor Tom Brokaw that were postmarked 18 September 2001. Those letters read:

THIS IS NEXT. TAKE PENACILIN NOW. DEATH TO
AMERICA. DEATH TO ISRAEL. ALLAH IS GREAT.

Besides having some sound medical advice in them, the powder laden with anthrax spores in those letters was far less pure than the ones mailed to Senators Daschle and Leahy in Washington DC, which resulted in the deaths of five people.

I have had a really hard time writing about the anthrax letters. Every now and again I take a walk through the Senate office buildings and try to imagine what the people who worked there at the time went through. People there don't speak about it, in much the way I don't speak easily about my experiences on September 11th. I suspect this is common amongst witnesses to tragedy.

To understand the reality of bioterrorism, we have to face the reality of what happened with the anthrax letters in unpleasant detail. Without such understanding, we are left with the now dulled panic caused by the attacks and the subsequent backwards rhetoric that fueled widespread misunderstanding of bioterrorism in the US.

Anthrax is a bacterium that spends most of its life in the soil as a benign spore. Occasionally, it infects cattle and even more rarely it will infect humans. Before the letters, the last human death from anthrax in the US was in 1976. The theory is that the man, who had not been near a farm, had breathed in spores that were on a wool sweater he had gotten from Afghanistan. This is a pretty far-fetched explanation, but plausible. Many bacteria can go into a dormancy stage where they transform into spores when conditions are not conducive to growth. Anthrax can stay in this state for decades. We actually don't know exactly how long they can stay dormant, but when it enters a growth phase it is for very brief periods of aggressive growth.

There are three main ways you can become infected with anthrax: coetaneous, gastrointestinal and inhalation. Coetaneous infection is by far the most common form and is caused by spores that gain access through a break or cut in the skin. This does not have to be a big cut, just a minuscule break in the skin. Theoretically you only need a single spore to cause an infection although spores usually come in clumps, so it is more likely that you'd get a bunch in.

Let's put it in Bush speak. 'Ya see, the spores are real small. Tiny. Can't see 'em. You put on yer glasses, still can't see 'em. It's really a faith thing. Ya have to believe they're there. Once they find a good ol' place to nestle in, they grow. It's hard work, but they do it. They change and grow and make a liaison.'

Too far? Once the spores enter a growth phase a skin lesion forms at the site of infection. The great majority of infections from the letters were coetaneous and were easily cured with antibiotics. Coetaneous anthrax infection is very rarely fatal, and most of the time will not progress further than the initial infection point.

Next is gastrointestinal infection, caused by eating food that is tainted with anthrax spores. This type of infection is very rare and is also cured pretty easily with antibiotics if caught in time.

Finally, there are inhalation anthrax infections – by far the deadliest, mostly because by the time doctors figure out what is going on, it is usually too late. The five people who died in 2001 all had inhalation anthrax. Once inhaled into the lungs, the immune system sends specialized cells called macrophages to swallow the spores, which develop into bacteria, which are then transported to the lymph nodes. At this point, the infected individual will start to show symptoms similar to the flu. In the lymph nodes they divide and make their way into the bloodstream, where they start to expel a toxin. Death comes quickly from toxic shock and organ failure. Inhalation anthrax can be treated with antibiotics, but once it is in the bloodstream it is often too late.

A lot has been said about the anthrax letters, some of it correct, some of it speculation. Let's play detective and see what we can figure out. First, the anthrax strain used was first isolated in Texas, but was named the Ames strain after Ames, Iowa, the location of the facility that identified it as pathogenic and unique. It has been used by the US biodefense program since it was identified as being very virulent.

A good way to understand what a bacterial or viral strain is is to think of horses. Horses are all the same species, and mostly look alike, but some are better suited for running and others are better suited for jumping. This has to do with their genetics. Bacteria are similar. Genetic changes that make one strain grow faster than another can mean the difference between a strain that is pathogenic and one that is relatively benign. You can think of them like twins. They look alike and they are genetically almost identical, but in the process of growing separately, small changes can happen that make them genetically distinguishable. When geneticists looked at the DNA from the anthrax found in the letters, they were able to identify it as the Ames strain. Some have estimated that there are as many as 50 or so labs that have this particular strain of anthrax, almost all of which are in the US. The truth is that it is really hard to tell who has the strain because it was distributed to so many places by the US government and the records were quite poor.

'Wait, wait, wait. Didn't the US sign the Biological and Toxin Weapons Convention of 1972 that bans our development of bioweapons like anthrax?' Excellent question. Yes we did, which is why, officially, the US does not have any biological weapons development programs and why we do not officially stockpile any biological or chemical weapons. But notice that we do not organize our military under the moniker 'War Department' any more either. We call it the Department of Defense. Under the convention, we can use the same logic to undertake any number of research projects to develop countermeasures to bioterror agents, and this means making bioterror agents that are at least as dangerous as our enemy has. The US therefore did develop biological weapons in the process of preparing for what our enemy might do, despite the fact that it meant that we had to develop bioweapons further than we had in 1972. These activities, while not necessarily in direct conflict with the convention, do go against the spirit of the deal. Also, the US has refused to sign on to an agreement to allow inspections of its biodefense facilities, which some have also said implies that we have not followed the convention to the letter.

This was most certainly the appropriate approach during the Cold War, since we know that the Soviet Union had a very sophisticated bioweapons program that concentrated on anthrax, smallpox and other agents. Their anthrax program has been well-documented by ex-Soviet scientists now working in the US and through a disastrous event in 1979 when anthrax was accidentally released from a Soviet military facility, resulting in at least 94 deaths. We also know of or suspect at least 18 other countries of having biological weapons or biodefense programs. Research into countermeasures to the agents being developed in these countries is a prudent strategy. The question is: do we need to make weaponized versions of an agent to make a vaccine or antibiotic to fight them? Most people in the drug-making business would say no. But to test vaccines and other countermeasures, you have to know how they will act, and thus many believe that weaponizing agents was and still is necessary.

Detailed information on our bioweapons program is of course classified, but we know that research on the Ames strain has been going on at Fort Detrick, Maryland, and at the US Army facility at the Dugway Proving Ground in Utah; some claim that the CIA has an advanced program. The Dugway Proving Ground is larger than the state of Rhode Island and has the facilities to make very high-grade weaponized anthrax. There, the military used to weaponize anthrax spores, ship them to Fort Detrick to be gamma irradiated so they were no longer dangerous, have the spores shipped back and use them in outdoor dispersal tests.

The FBI initially thought that it was likely that a single US-trained scientist or person with some medical training was likely behind the anthrax attacks. While this is certainly a possibility, based on what we know it could have been more than one person involved in making the powder, or one very skilled biological weapons maker. Since the letters were such a significant series of events in the history of bioterror, I think it's important to understand the basics of making weaponized anthrax. Also, understanding what is involved in weaponizing anthrax could give us significant clues about who was responsible for the letters.

To weaponize anthrax, you have to take the following steps. First, you have to acquire the strain. It would not have been trivial to get hold of the specific Ames strain, but the letter sender might have had access without doing anything out of their normal routine. We are talking about one of the most dangerous bacterial strains in the world, but nonetheless, many labs have it. Next you have to grow a bunch of it, first on a culture plate and then in a flask with special nutrient rich liquid that is optimized for anthrax growth. If you do it wrong, other bacteria will grow in the liquid. Then you have to starve the anthrax so it turns into a spore. In nature, anthrax turns into spores when conditions are too harsh for it to grow. At this point you will have a mixture of dead anthrax cells and spores that could cause coetaneous anthrax but would not aerosolize well. The process is easy enough to do with a little more training and there are thousands of people who could easily learn to do this. Not just any schlub off the street can turn bacteria into a spore; in fact, most PhD level biologists wouldn't have a clue as to the specifics of doing this for anthrax.

According to congressional testimony by scientists who examined electron micrographs of the spores from some of the letters and several other sources, the first letters, mailed to newsrooms in New York, did not have a high concentration of spores in the powder. They did have bits of dead bacterial cells and the sizes of the clumps of spores were larger and more varied than the letters mailed to Senators Daschle and Leahy. That indicates that the person(s) had access to anthrax at different stages of preparation or prepared multiple batches. The letters sent to the Senate had an extraordinarily high concentration of spores and with smaller clumps of spores, which are more easily aerosolized. This is an important fact. It is thought that the spores in the second batch were concentrated enough and small enough so they could make their way through the pores in the paper in the letter and envelope, or were just forced out and easily aerosolized as the letters shot through a mail sorter.

To purify anthrax spores, you have to mix them in liquid and centrifuge them to get rid of all the larger dead bacteria parts. This might have to be repeated to get the spores really pure. This takes some skill in the lab as well and significant know-how about spore purification. Next you dry the anthrax and weaponize it. Two basic ways to do it have been described: milling and spray drying. The purpose of both of these techniques is to make smaller particles containing fewer spores or (better) single spores. Milling spores involves drying out the purified spores and grinding them up to make smaller particles. Spray drying involves mixing the spores in liquid which is sprayed into an enclosed tube and blasted with hot air. The hot air evaporates the liquid, leaving spores.

There were some controversial reports that the spores in the Senate letters were coated with silica. The reason that additives like silica are used is simple. The silica acts like little ball bearings that reduce the tendency of spores to stick to each other; in other words, it reduces clumping. The silica can also take on a static charge, which can repel the spores from each other, further reducing clumping. Finally, coating agents like silica can absorb some water and keep the spores dry. This class of additives has but one purpose in microbial research: weaponizing agents. They are not used to produce aerosolized medications that are designed to be inhaled because the body cannot easily clear the silica from the lungs. In 2006, the FBI revealed in a research paper that detailed how they did their initial forensic investigation that the anthrax had not been treated with silica. What they did not reveal was why it took them until July of 2002 (10 months after the attacks) to find the mailbox it was mailed from. There were only 628 mailboxes in the area where the letter was mailed and there were still spores in the mailbox when they found it. Nonetheless, when you consider the number of labs that would have had the Ames strain at the time and persons who received an anthrax vaccine, which would certainly have been required, you still have a very narrow field.

The anthrax research community used to be quite small relative to other forms of research. Someone out there worked with or trained the person who sent the letters or works in a facility that allowed someone access to a dangerous weapon.

One reason the FBI believes that the attacks were likely done by a single American scientist is that the person probably did not intend a lot of people to die. Sounds crazy – right? Think about it. The person warned about the contents of the envelope in the letter. If this person had the expertise to handle anthrax, then they also knew that once a victim is warned that they have just been exposed, they can take prophylactic antibiotics and won't die. Also, putting it in a letter is a pretty ineffective way of spreading inhalation anthrax, as seen by the relatively low number of people that were infected. The letters were folded in a way consistent with someone who tried to contain the powder, a so-called pharmacist's fold. If the attacker had just sprinkled the powder into the mailboxes, hundreds or thousands more people would have been infected with inhalation anthrax because the spores would not have had to work their way through layers of paper to become airborne. The mail carrier would not have given the odd powder that was coating the letters much thought, and a much larger spread of spores and death across the country would certainly have occurred. This person wanted to wield power and wanted to scare people. It just doesn't make sense that they planned for spores to be filtered through the paper of the letter and envelope and be aerosolized. Whoever did it was a sociopath, but not stupid. They just didn't consider the leakage issue.

The wording in the letters implied that the person who sent them was from the Middle East and was Muslim. But that doesn't make any sense either. The Washington letters were addressed to quite liberal Democrat Senators. If the person was mad at the US government, why would they send them to two men who don't have a history of making reckless statements about Arab countries (as the President and several Republican congressmen have) and did not advocate force in dealing with international diplomatic situations. One might think that the letter writer, if he is indeed American, has a thing against Democrats.

It's a grim suggestion, but one that is not getting the kind of play in the press that it should. Most acts of terrorism are home-grown. Even groups like Al Qaeda have recognized that locally organized terrorism is harder to detect. The train bombings in Madrid and London were organized and executed by local terrorists, not people who immigrated to the country they were trying to attack. The attacks on the Pentagon and the two attacks on the World Trade Center, 1993 and 2001, were unusual in the terrorism world. Here, domestic terrorist acts are traditionally born right out of the hate and dismay we have right here in the great ol' US of A, the most notorious being Ted Kaczynski, Timothy McVeigh and Eric Rudolph. The same holds true for attacks in other countries. The combined evidence – access, weaponizing expertise, targets, a desire to scare but not kill people, language meant to redirect hatred towards Arabs – puts the focus on a military-trained scientist, and that is where the FBI started their investigation. This most certainly does not mean that the person wore a uniform. Most military facilities that handle anthrax employ many civilian staff, so this misconception ought to be rejected out of hand.

That the letters were mailed from New Jersey implies that the mailer could have been working at the Fort Detrick facility, a short drive down the turnpike. But it is also possible that the mailer deliberately mailed the letters in close proximity to Fort Detrick and that he was not the person who made the anthrax. There have been many reports that the US military biological weapons labs have had lax security, sloppy bookkeeping and some serious job satisfaction issues. The investigation by the military and FBI got very quiet very quickly when the press started to piece together the puzzle.

At one point, the finger was pointed in the press and through officials associated with the investigation at one particular man who worked at Fort Detrick. The FBI called him a 'person of interest.' As it turns out, this man was probably wrongly accused. His career and life were dragged through the mud like the security guard initially accused of the Olympic bombing in Atlanta. (I won't give either of their names here.) The FBI apparently hounded this man relentlessly and he has filed suit against former Attorney General John Ashcroft and the Justice Department. Behind the scenes in the biosecurity community, a select few pointed the finger at him in the press, mostly without actually naming him. The whole situation makes me sick. It's as if we have learned nothing.

The FBI and the DoD have not been open with the American public about the investigation. Not the details, just the progress. The *Washington Post* revealed in September 2006 that the FBI had reduced the number of agents on the case from 31 to 17. They have issued over 5,000 subpoenas, interviewed 9,100 people, and chased leads on four continents. Yet with all those efforts, they remain interested in Fort Detrick, the Proving Grounds and Louisiana State University. They even drained a pond looking for evidence. If we put aside the conspiracy theories, then we are forced to accept that we are living with someone in our midst who has the potential to kill thousands upon thousands of people. Maybe it's time to talk more candidly about the biodefense work that went on in the US. In the end, I think we can count on the public being more interested in Janet Jackson's breast and the *American Idol* finals than our biodefense programs.

Since the initial anthrax attacks in 2001 there have been tens of thousands of fake anthrax letters mailed in attempts to scare the hell out of people. These letters have been filled with white powders ranging from powdered sugar to rat poison, but we have had no news on the suspect or suspects who killed five people, disrupted the lives of thousands and attacked the United States Congress.

Homeland absurdity

The US State Department defines terrorism (yes, they had meetings to discuss the wording of the definition of terrorism; no bureaucracy there) as:

> Premeditated, politically motivated violence perpetrated against noncombatant targets by sub-national groups or clandestine agents, usually intended to influence an audience.

The FBI, it seems, also had a powwow or two to define terrorism. They define it as:

> The unlawful use of force and violence against persons or property to intimidate or coerce a government, the civilian population, or any segment thereof, in furtherance of political or social objectives.

While ponderous, these are catch-all definitions, the need for which is precisely the point: people will take up bioterror and other forms of terror for any number of reasons. Since it is outside of state-run programs and does not use conventional military actions, it is usually dramatic and small in scale compared with waging war.

Unfortunately, the Bush administration has gone to old Cold War scare tactics in a sad display of naiveté and confidence in the knee-jerk mentality and ignorance of the American people. Under former Secretary Tom Ridge, the Department of Homeland Security introduced a laughable plan (see box) for dealing with terrorist attacks that drew a lot of bad press and jokes on late night television (jokes about the color-coded threat levels still persist). It used to be detailed on the Department of Homeland Security's website (it was not

1. Assemble an emergency kit

All of us should be able to survive comfortably on our own for at least a three-day period. That's the amount of time you may need to remain in your home until the danger from a biological, chemical or radiological attack has passed. You'll need:

- A change of clothes
- Sleeping bags
- Food and water. A gallon of water per person per day should be enough. Canned and dried foods are easy to store and prepare.

Our advice is to start now by gathering basic emergency supplies – a flashlight, a battery-powered radio, extra batteries, a first-aid kit, prescription medicines and toilet articles. Duct tape and heavy-duty plastic garbage bags can be used to seal windows and doors. Make sure all household members know where the kit is kept. You should also consider bringing a disaster supply kit to work or leaving one in your car.

2. Make a family communication plan

- Your family may not be together at home when an attack occurs. Make sure everyone knows contact numbers and how to get in touch.
- It may be wise to have everyone call an out-of-state friend or relative.
- Keep a list of emergency numbers near the phone.
- Select a 'safe-room' where everyone can gather. The best choice is an interior room above ground with few windows and doors.

3. Learn more about readiness

Planning helps. If your family knows what to expect, they will be calmer in the aftermath of a terrorist event. For example, you should find out where to turn for instructions, such as local broadcasting networks. Local authorities will broadcast information as quickly as possible concerning the nature of the emergency and what you should do next. Be sure to keep listening for updates.

There are other ways to plan ahead. Take a first aid and CPR class so that you can provide emergency medical help. Review your insurance policies to reduce the economic impact of a potential disaster. Remember to make accomodations for elderly family members and neighbors or those with special needs. Finally, try to make arrangements for pets not allowed in public shelters.

replaced until mid-2005) and includes creating a plan of action for your family to wall themselves from the environment for at least three days.

Ponderous! I particularly enjoy the misspelling of 'accommodations.' It is a national embarrassment that underscores the administration's belief that if you exaggerate and keep people scared, they will follow without question. Unfortunately they are mostly right.

Since we know that we would not likely be aware of a biological attack until all or most primary infections are well under way, we can conclude that duct tape and plastic will not save lives in the case of a bioterror attack. It will only give people the impression that they are doing something proactive – empowering them when they are powerless. It is important to point out that if an agent – anthrax spores for example – has already entered the home, duct taping yourself in will basically ensure that you lock in all that anthraxy goodness and ensure infection. Yummy!

The three-day estimate for clearance of agents is a complete mystery and the expectation that people will come up with – much less follow – a family plan when they are being told that the country is under attack is as improbable as it is picayune. The Department of Homeland Security is so concerned with not looking foolish and unprepared in the face of a real attack that it raised the threat level over and over again to the point that the public doesn't take any of their warnings seriously. It was particularly interesting that the threat level was raised frequently before the 2004 election, but not too often since then. The plan does establish a baseline level of government stupidity that we can build from though. To best understand and cope with bioterrorism, government and the public have to have a good handle on what to expect to prevent paranoia, panic and needless overreaction.

There are seven basic areas for building public health capacity and preparedness for bioterror and other public health emergencies.

1. You need to train a workforce of health professionals to recognize, test, treat and possibly quarantine people who have been exposed to bioterror agents.
2. You have to establish significant lab capacity to test samples, have adequate treatment capacity for distribution of vaccines and drugs, and quarantine capacity for those who have been exposed to an infectious agent.
3. You need centralized epidemiology and disease surveillance capability that has the ability to coordinate with local health officials to identify an attack quickly.
4. There must be secure information systems for coordinated communication to responders in the event of an attack.
5. There must be defined coordinated systems to communicate with the public so they can be advised on a proper course of action and avoid misinformation.
6. There has to be a way to oversee preparedness on all levels and properly evaluate the level of preparedness and action that can be taken in the event of an attack.
7. There have to be well-defined coordinated plans of action on the local and federal levels including detailed protocols for individual responders and agencies that can be evaluated for effectiveness and flexibility.

They have not done this and we currently are not fully prepared in any of these areas.

Under the circumstances, it is understandable that Congress was upset. They had just been attacked with biological weapons. But in the rush to respond to the threat of bioterror, to quash the possibility of this ever happening again, in 2002 Project Bioshield passed Congress. The price tag to taxpayers was $5.6 billion and the project was aimed at addressing the holes in our infrastructure that could have been exploited to make the attack far worse. This is pitifully little money to throw at a problem of decay, but it was a good first step. Of particular concern is that throwing money at the public health infrastructure often strengthened the existing divisions between federal, state and local health agencies; cooperation was not stressed enough.

One must understand that the divisions were formed out of under-funded bureaucracies and that any capacity building would be a welcome change. All preliminary progress reports indicate that we are a long way away from being able to deal with a large public health emergency. Unfortunately, much of the money has been spent on preparing for unlikely bioterror-specific scenarios – indiscriminate spending of money that doesn't seriously address preparedness or get us any closer to being ready to respond to an emergency, while simultaneously ignoring the public health crises which are more likely to arise.

Unfortunately, the system put in place was tested in 2005 when Hurricane Katrina destroyed the gulf coast of Alabama, Mississippi and Louisiana. Despite the fact that an emergency was declared in the city of New Orleans before the storm even hit, no significant help was sent for days. If there was ever an impeachable offense, it is the lack of preparedness and lack of response by President Bush to this disaster. While the blame for the lack of response falls firmly on his shoulders, he had plenty of help in failing the country from Congress.

The best way to view the attitude of politicians toward homeland security is to look at how they dole out 'first responders' grants to different states. A formula was set up through which every state received 0.75% of the money from the largest pools of homeland security funds. Adding it up, you realize that the US spent over one third of the money without looking at the overall need of the country. Like hogs at the trough, politicians ensured that Newark, New Jersey has air-conditioned garbage trucks; Converse, Texas has a trailer that they use to transport riding lawn mowers to lawnmower races; Columbus, Ohio has bulletproof vests for dogs; and Washington DC cops have some swanky leather jackets and the sanitation workers got to go to a Dale Carne-

gie course. But that doesn't top the city's $100,000 Home-
land Security *rap song!* It's disturbing that Wyoming got the
highest per capita amount of Homeland Security money.
(Note to terrorists: there must be some really good stuff out
in Wyoming if they are spending so much on protecting it.
After all, Dick Cheney is from Wyoming. New York had a very
low level of funding per capita; nothing good there.) Spending bills for the Department of Homeland Security have been
chock full of pork spending as well. No wonder there is not a
single major city in the US that is prepared for a bioterror
attack, much less any public health disaster.

In June 2006, it was revealed that when the Department of
Homeland Security was evaluating New York City for its
share of anti-terrorism funding that it did not recognize the
city as having any icons or monuments that could be targets
for terrorism. Not surprisingly, they decided to cut the city's
funding by 40%. In the same evaluation, they found that
Indiana had twice as many potential targets as California,
that a petting zoo in Woodville, Alabama and a flea market in
Sweetwater, Tennessee were considered potential terror targets. Let's consider some of the other targets, shall we? An
ice cream parlor is a potential terrorism target, but Times
Square is not. The Amish Country Popcorn Factory is a
target, but the Empire State Building is definitely not. The
Apple and Pork Festival in Clinton, Illinois is a primo target,
but Wall Street is so 2001 as far as targets go. Makes sense to
me.

While Congress was feeding its special interests with
homeland security money, it failed to provide a sound version of the bioshield legislation, strong enough to entice
pharmaceutical companies to start making drugs to fight
bioterrror agents. The main reason is that these drugs would
not have a large market. For example, antibiotics and vaccines are far less profitable for pharmaceutical companies
simply because they cure or prevent disease. Logically, developing a drug for high cholesterol, hypertension or some

other chronic condition requires that a patient be on the drug for a long time and there would be more people taking it, and thus more profit. The number of companies working on vaccines and antibiotics decreased dramatically during the 1990s for this very reason. It is more profitable to come out with something that innately has a larger market.

So Congress decided that it needed to ladle on more incentives for bioterrorism countermeasures. The issue here is motive; the pharmaceutical companies are primarily concerned with making money, and this is to be expected. Like all companies, they have investors to answer to. Goodwill goes a long way, but investors would much rather see a strong quarterly earnings report than hear that drugs were donated to families in Africa or that they were developing countermeasures to an unknown bioterror threat. In 2004, a study came out that detailed that of the 506 drugs in late stage clinical testing by the world's top 15 pharmaceutical companies, there were only six antibacterials, all of which were derivations of already established antibiotics. There were four drugs being developed for erectile dysfunction though. I suppose there are many ways to consider benevolence, but it should be pointed out that even if we don't see a bacteria-based bioterror attack, the CDC estimates that about 90,000 Americans will die this year from infections they acquire in hospitals and about 70% of those infections will be from bacteria that have evolved resistance to at least one antibiotic.

But the pharmaceutical industry is hardly to blame for the lack of countermeasure development. The administration has not provided them with a priority list for antibiotics, antivirals, vaccines and detection tests. Without a clear statement about what countermeasures the government is interested in stockpiling, or which they will purchase for active public health preparedness, why in the world would they develop them? There is no cohesive plan and the administration has absolutely failed. They originally awarded a $900 million contract to produce 75 million doses of a next generation anthrax vaccine to a company that had never produced a product. Not unexpectedly, the company failed to meet their obligation and the government cancelled the contract and ordered more of the old vaccine, which is delivered over many months and has some rather nasty side-effects. The government has still not said that they would buy other countermeasures or drugs. How is a pharmaceutical company supposed to invest in research and development without a market or with the threat that their contract will be cancelled? It would be absolutely stupid for them to get involved.

More recently, Congress tried to pass legislation aimed at providing pharmaceutical companies large incentives for developing any countermeasures. In one proposal that was *seriously* being considered, if a company came up with an effective new antibiotic or vaccine aimed at fighting a potential bioterror attack, it would have earned the ability to extend the patents on any other drug they had on the market. So, in effect, if they developed a countermeasure, it would have prevented generic forms of common drugs from making it to the market, costing Americans billions of dollars in additional healthcare costs. This was affectionately called the 'Wild Card' provision, and caused quite a bit of controversy from consumer advocates and generic drug companies. Basically, it would have been a tax on people who need to take prescription medication.

Much of the bioweapon countermeasure research through the Department of Defense has focused on vaccine development. However, the DoD has not sought FDA approval of these vaccines for use in the general public and concern has been raised about the number of vaccines that are being given to troops. One DoD official implied recently that he considered the FDA to be somewhat bothersome and that the high standards they have for vaccines might be too high. This is perhaps true, but nonetheless the DoD has not set up its research funding structure to enable them to afford the rigorous FDA vaccine approval process. I think most people would agree that the lives and health of troops should never be put in danger through the use of vaccines that have not been put through the most rigorous of trials.

In addition, we have not developed rapid tests, on the order of minutes, for most biological agents thought to be bioterror threats. I am talking about technology akin to that used in home pregnancy tests, which detects minute amounts of protein quickly. One really has to question where the DoD biodefense budget has gone if they have not developed a full complement of effective countermeasures and detection reagents to a known set of threat agents.

The military do many things very well, but biology research is not one of them. They just don't have the training infrastructure equivalent to any of the top research facilities in the world and on the whole are not drawing in the best and brightest scientists into their facilities. This is not to say that there are no bright, talented people working in military facilities. They just aren't the best of the best.

If we consider the FBI's theory that it was a lone domestic scientist that sent the letters, then one might logically conclude that in the future we will probably want to keep close tabs on anyone who might be trained well enough to access, grow, weaponize and disseminate bioweapons, and limit the number of people doing research on the most serious of bioterror agents. The US response, however, has been to dramatically expand the research infrastructure for bioterror agents at an unprecedented rate. Most of that research is not aimed at developing simple countermeasures like antibiotics and antivirals.

Safety first

There are specific guidelines for working with pathogenic microorganisms and animals infected with microorganisms. These guidelines are delineated into a series of biosafety levels that increase in number based on the level of danger that the pathogen poses to people, animals and the environment as well as the level of protective equipment and procedures needed by researchers.

Biosafety Level 1 (BSL-1) is designated for agents not known consistently to cause human disease and does not require any special equipment. Level 2 is for pathogens associated with human disease and requires a series of simple protocols for decontamination: face protection (usually in the form of a fume hood), gloves and the ability to physically contain the agent. Measles, salmonella and hepatitis B are hard to get unless ingested or injected so they make the BSL-2 list.

BSL-3 is designated for indigenous or exotic agents with potential for aerosol transmission where the disease may have serious consequences – tuberculosis and plague for example. It requires respiratory protection, a negative airflow, exhausted air (not recirculated), double door access, and physical separation from other lab space. Pretty serious stuff.

BSL-4 is the king. It is reserved for agents that pose a high risk of life-threatening disease or aerosol-transmitted lab infections. It requires people working with them to shower in and out of the lab, a separate building, and full body suits with positive air pressure. Agents like Ebola virus are on this list. These are the agents for which if you get infected, you are pretty screwed.

After the anthrax letters, the US expanded research on potential bioterror agents to an unprecedented level. Part of that expansion included building at least nine new BSL-3 labs funded by the National Institute of Allergy and Infectious Disease. It is not known how much of an increase this represents above and beyond the capacity that already existed because there is no accurate account of the BSL-3 capacity that existed beforehand. Some estimates place it at over 600 separate facilities nationwide. Also, there is no assessment of how much BSL-3 lab space is being built in pri-

Citizen Butler

Dr Thomas Butler is one of the preeminent infectious disease researchers in the US. The WHO credits his work, which helped lead to a treatment for cholera, with saving the lives of over 2 million people. In 2003, Dr Butler, then at Texas Tech University, was convicted of charges related to an investigation of 30 missing samples of plague bacteria. Earlier that year, Butler reported that the samples were missing from his lab, which led to Attorney General John Ashcroft (reportedly briefing the President and Homeland Security Chief Tom Ridge) to say that 'We may have a plague attack in Texas.' This resulted in 60 FBI agents heading to Lubbock, Texas to investigate Butler. Under duress after hours of interrogation without legal counsel, Dr Butler signed a statement saying that he had previously destroyed the vials. Dr Butler was led to believe that the statement would defuse the circus atmosphere that had been created by the overzealous investigation, but he was arrested instead. When Dr Butler refused to take a plea bargain on charges of lying to the FBI and illegally transporting plague, they ladled on 54 more charges as further punishment, with a total of a possible 469 years in prison and $17 million dollars in fines. In the end, Dr Butler was convicted of not checking the proper box on a FedEx shipping label and mishandling funds in his lab, but not on any of the terrorism-related or lying charges.

This type of prosecution does not protect the American people and has prevented a good man from doing his work. Dr Butler has had his career destroyed and his family devastated for the vanity of the Bush administration, especially former Attorney General John Ashcroft. They actually count this man's conviction as a victory in the war on terror. The scientific community, including the National Academies of Science, disagrees. Dr. Butler was released from jail in early 2006 and is still trying to put his life back together and clear his name. (For further developments visit http://www.fas.org/butler/).

vate institutes. There was also a massive 11,000 square meters of BSL-4 lab space being built. Before 2001, there was roughly 1,700 square meters of BSL-4 space in the entire country, not counting government lab space that is not accessible to public researchers. This unprecedented build-up in lab capacity comes at a price though. When the government built capacity with no plan in mind, it neglected to consider that there would be a massive need to train and support researchers to work in the space for a long period of time, which would eventually start to detract from other research priorities.

The new focus has come under fire as the funding structure of NIH has shifted to dealing with BSL-4 and BSL-3 agents over agents that cause more common health problems, kill more people and pose a greater national health risk. That shift resulted in grants for people working on common microbiology organisms dropping. One estimate was that there was a 2,194% increase in biodefense funding for the National Institute of Allergy and Infectious Disease, but only a 62.4% increase in overall funding for the institute. There is some controversy over the extent of the funding drop for non-bioterror microbial research (depending upon which grants you count), but it is estimated by some to be somewhere between 26% and 41%. Last year, the NIH agreed to review the funding shift, but in the end, once you build enormous capacity, you have to fill the space. Non-bioterror microbial research will continue to take the hit for the reactionary spending spree. But building capacity does not guarantee safety.

In 2005, it was revealed that an instrument used to infect animals with aerosolized BSL-3 agents was responsible for three Seattle researchers becoming infected with tuberculosis. The chamber, called a Madison chamber after the University of Wisconsin, is approved for BSL-3 work and has been touted by its inventor as 'foolproof'; he claimed that 'respirator use was not necessary' for researchers using the device. Tuberculosis is highly contagious and could have resulted in far more infections had they not been discovered during routine testing. The infections actually took place in 2004, but were not reported until several months later. The company that made the device was fined $2000. The chamber was still in use in at least nine states and several countries at the time I wrote this. In addition, in September 2005 it was revealed that three mice infected with plague were missing from an animal disease research facility at the Public Health Research Institute in New Jersey. While it is possible that the mice escaped, were eaten by other mice, were stolen or there was a paperwork error, rest assured: they would have died of plague within days.

Animals and equipment are not the only issue. In 2004, three Boston University researchers were infected with a highly pathogenic strain of bacteria called *Tularemia*, which causes rabbit fever – one in May, and two more in September. But the disease was not identified as tularemia until October. They were supposed to be working with a non-disease-causing strain, but somehow wound up working on the dangerous one. This is actually the second time that researchers at the university had been infected with tularemia, underscoring the inherent danger of working on pathogenic agents even when trained to do so. What if this human error had involved a highly transmissible virus? I am comforted by the fact that there are no documented secondary infections of people in the US with any research agent, but this is not the case everywhere. There were several cases in Asia where people who were not working with SARS got it from infected colleagues who did work with it.

Unfortunately, these are only three examples of a series of reports of mishandling of infectious agents including several reports out of the US Army Medical Research Institute of Infectious Diseases at Fort Detrick. Remember: this is the facility that is getting more BSL-4 space, and that the FBI thinks could be the source of the anthrax used in the 2001 letters.

Influenza is another good example. It has been categorized as a BSL-2 pathogen for years. It is generally not a dangerous virus to work with in the lab because standard laboratory practices prevent researchers from getting infected. Researchers working with influenza also generally get immunized regularly. However, pandemic strains of influenza are stored in research labs around the world. Recently, the CDC upped the ante and now recommends that people working with pandemic influenza strains work in a BSL-3 facility.

Unfortunately, safety is not just a concern in the research lab, it is a serious concern when it comes to vaccines and drugs used as countermeasures. Smallpox is often spoken of as a serious bioterror threat. There are a few origins to this idea, starting with the Soviet smallpox program, which produced vats full of weapons grade virus. Once the Soviet Union fell apart, many of the scientists involved with the program were out of work. It is feared that they started working for other states with less than progressive attitudes towards the use of bioweapons. Smallpox would certainly be an effective bioweapon, as about 30% of people who contract it die. Several Native American tribes were virtually exterminated by the virus. Naturally occurring smallpox has not been seen in the US since 1949 and the last case reported in the world occurred in Somalia in 1977. The US stopped vaccinating people for smallpox in 1972 and the disease was declared officially eradicated from the world in 1980. Nonetheless, in 2002 the Bush administration announced that it wanted to vaccinate 450,000 public health workers as well as active duty military personnel headed for Iraq or Afghanistan for smallpox, and in addition wanted to stockpile the vaccine around the country. But, much to the frustration of the CDC and the Department of Health and Human Services, no explanation was provided for the new vaccine program.

The smallpox vaccine itself is a live virus called vaccinia virus that is related to smallpox virus. Vaccinia virus is contagious, so someone who is vaccinated can actually pass the virus onto someone else who is in contact with fluids from the site of the injection. Also, the vaccine itself can have some pretty nasty side-effects, which are characterized as potentially life threatening in somewhere between 15 and 50 out of every 1 million people vaccinated. At least one person is known to have died as a direct result of the President's vaccination program. Needless to say, the voluntary program was a total failure. Without justification, people just didn't want to risk their lives. The implication was that the administration knew something about a credible smallpox threat to the US. After having been lied to about many aspects of Iraq's weapons of mass destruction programs, it was of little surprise that people didn't jump at the opportunity to get vaccinated.

I find it hard to believe that a President would call for the vaccination of over a million people if there was no known threat. Otherwise, this would be one of the most irresponsible domestic health requests that a US President has ever made. So, why not be upfront with the public about a credible threat? There are a couple of reasons, but let's explore one. Let's say a Pakistani government vehicle carrying a few barrels of smallpox was stopped by American soldiers as it attempted to enter Afghanistan. Well, since Pakistan is our partner in the 'global war on terror,' it would be a total diplomatic disaster to tell the American people that our ally was shipping bioweapons across the border to a country we are fighting a war in. That's just one conspiracy theory scenario, but you get the picture.

There has been a lot of chatter about the possibility of a large bioterror attack in the US. Most of those scenarios involve either aerosolized anthrax or smallpox. But there is one real possibility that has not gotten significant play: the flu. I know; it's not as scary as anthrax. It doesn't even have a cool name. But approximately 36,000 people die each year from the flu in the US alone, many more if you factor in the secondary effects like cardiac damage caused by the infection. In the last century there were several pandemic infections of highly virulent flu strains that killed millions of people. Take the 1918 Spanish flu pandemic that spread around the world very quickly, for example, and killed an estimated 20–50 million people worldwide, including 500,000 Americans. This flu only killed around 1–2% of

Coming to a station near you

The biggest chemical attack threat is not from people making agents and dispersing them. I've already mentioned the issue with a chemical or biological attack on our water supply and the ease of inactivating the chlorine in it, but how do you think chlorine gets to those plants? By truck and rail. It is common for railcars full of liquid chlorine or other highly dangerous chemicals to pass right through the center of a city. This is actually the case for Washington DC. Trains carrying chlorine travel a couple of blocks from the Capitol building.

Besides being good for killing algae and bacteria in your pool and drinking water, chlorine can be very toxic to humans. Much of what we know about dealing with chemical disasters actually comes from dealing with chlorine spills in communities. This is one of those threats that we have known about for decades and have done almost nothing to stop. A significant explosion under one of those cars could cause it to rupture and leak. The liquid would quickly evaporate and become deadly chlorine gas. It is estimated that a single tanker car full of chlorine, if spilled would kill or injure over 100,000 people in Washington DC, could send a toxic gas cloud that would cover over 40 miles and as it traveled by wind would create an area of death that would extend for 4 miles by 14.5 miles. Yet, rail cars carrying chlorine and other dangerous chemicals are allowed to travel right through major cities with no real security. Usually, these trains are not even destined for the city; they are just passing through. The Justice Department actually obtained a court order preventing the District of Columbia from enacting a law to prevent the practice. What's more, the Department of Homeland Security has proposed installing cameras in high-risk areas rather than recommending that they be rerouted around cities unless destined for them. I guess the cameras would be useful for catching the terrorists, if they looked right into them and smiled after they kill thousands of people, because it is highly unlikely that this would stop them if they wished to pull off such an act. In addition, there are several areas of the US that are chock-full of chemical plants that are completely vulnerable to terrorism. In New Jersey, a few miles from New York City, there is a corridor of chemical plants that have been shown to be completely vulnerable by journalists who have ventured to test their security.

those infected, but since everyone eventually gets the flu, that was enough. This came at a time when the world population was about one third of what it is today. So an agent does not have to kill a high percentage of people who are infected to be a serious threat. In fact, there have been several deadly pandemics, including 1957 Asian flu and 1968 Hong Kong flu.

In 2005, it was revealed that the strain of influenza that caused the 1957 pandemic in Asia was mistakenly mailed out in kits to close to 4,000 labs across the world, including labs in Saudi Arabia and Lebanon. The strain was to be used as a reference for researchers trying to identify markers for specific strains of influenza. When the error was discovered, there was a rapid call for labs to destroy the strain. However, even when the tubes were destroyed, there is no guarantee that someone in the lab didn't take a little sample to hold on to for future experiments.

There is also serious concern over some current research on the 1918 pandemic flu strain. Until recently, there were no samples of this virus around to study, but the discovery of frozen tissue from people who died of the 1918 flu gave some virologists the idea of remaking it. This caused tremendous controversy among some scientists, who see no real purpose in reconstituting a deadly virus that is not a current health threat anywhere in the world.

However, there is both intellectual and health value in determining what it is about that particular flu strain that was so deadly. It is not as if the scientists involved did this work in secret either: they had permission to reconstitute the virus from both the NIH and CDC. While there is plenty of research going on that I believe has questionable value, I welcome a central idea of basic science research: that you never know where a fundamental advance will come. Previous work on parts of the virus had already shown much about why this virus was deadly and the reconstitution of the virus also revealed important information.

Just for the record, it is highly unlikely that the 1918 flu will ever be effectively used as a bioterror agent because we have some immunity against it from being exposed to similar strains.

Like the planes that were flown into the World Trade Center and Pentagon, like the bombing in Oklahoma City, like the 1993 bombing of the World Trade Center, like the school shootings in Columbine, Colorado, like the Unabomber and countless other terrorist attacks around the world, the anthrax letters spelled out in no uncertain terms that if maniacs have the means to create chaos and destruction, there is little we can do to prevent it. As soon as we lock down one means of destruction, those with the will to hurt others will find another way.

Unfortunately, the new efforts are an extension of the misguided war on terror. Misinformation and a gangly uncoordinated public health system are our biggest vulnerabilities, not just to bioterrorism, but to all forms of terrorism and public health emergencies. Unfortunately, throwing money at the issue will only partially fix the problem. We are in serious need of leadership and public education that leads to organization and efficiency in our health care system. More people died of infections acquired in hospitals, the common cold, obesity-induced heart disease or cigarette smoking last year than ever died from bioterrorism, but we don't panic about these problems. The massive amount of money spent on terrorism preparedness has left us scarcely prepared to deal with an attack. As a nation, we have to take our collective heads out of our asses.

and another thing: the coming pandemic

We're all gonna *die*! Just kidding. Truth is that we can no better predict when an avian flu pandemic will start than we can predict when and where lightning will strike. We know what conditions are necessary, and what things can increase the risk of a strike, but would be hard pressed to give any details. So what's the deal?

Right now, three types of flu virus are circulating in the human population: H1N1, H1N2 and H3N2. Flu viruses are named after variants of two proteins that stick out from their surfaces like spikes: hemagglutinin and neuraminidase. The hemagglutinin spike allows a virus to enter cells. The neuraminidase spike enables it to get out again, having used the cell's molecular machinery to copy itself, destroying the cell in the process. These two proteins are what the immune system recognizes when a flu virus particle enters the body.

Every once in a while a new flu virus crops up and makes its way into humans. Since we don't have immunity to new viruses we are particularly susceptible to them. New viruses arise because in addition to infecting humans, influenza viruses infect birds and other animals, including pigs. A process of genetic reshuffling occurs very rarely when two different flu strains infect the same cell, usually in a pig or migratory bird. If the two are significantly different, their exchange of different hemagglutinin and neuraminidase variants can create a deadly new virus. Every once in a while one of these new flu strains crosses over to humans. Humans have little defense against these new strains and thus anyone who gets it is in grave danger.

Today's panic is over a new influenza strain: H5N1, aka 'avian flu'. At first, this was only found in wild birds, but it started to infect poultry in the markets where birds are kept crammed in cages. As of fall 2005, only about 120 people, who were in direct contact with infected birds, had been infected. About half of them died. Person to person transmission of the virus is only suspected in a few cases where people caught it while caring for a family member. Due to the high lethality to humans and birds, hundreds of millions of poultry throughout Asia have been killed in an attempt to control the spread of the virus. This will not *stop* the virus from spreading, but the logic was that it could *slow* the spread significantly. It didn't work. This is because viruses spread very easily in poultry since they are farmed in extremely cramped and unsanitary conditions and wild birds often mingle with them. More importantly, the wild birds are the ones spreading the virus, not the chickens. Duh!

But H5N1 is not just spreading in birds and other animals; it is also mutating. This is the start of a process of adaptation for the virus. Influenza viruses do not proofread their genome when they replicate, so they acquire mutations faster than animals or even other types of virus, which is precisely why the flu vaccine must be updated every year. That adaptation appears to have continued, as there are now birds and people who are catching this flu who are not dying, indicating that it is becoming less virulent. On the surface, this sounds like a good thing, but as viruses go this is actually a very bad sign. It indicates that the virus could be on its way to acquiring the ability to pass more easily from human to human. Even though a lower percentage of infected people would die from it, it would spread very quickly. The most deadly influenza strain on record, the 1918 flu that killed between 20 and 50 million people worldwide, only killed 1–2% of people infected, so lower virulence is not necessarily good news. Even if only 5 or 10% of people infected with the new flu died from it, you could get tens of millions of deaths. Scientists are now scrambling to come up with a vaccine for avian flu in fear of a pandemic, but since they do not have a good idea of what it will look like genetically if it acquires the ability to pass from human to human, there is no telling how effective the vaccine will be.

No one can accurately predict if or when there will be a flu pandemic. We know that the initial genetic shift that creates the possibility of a pandemic has happened, but we cannot predict if and when enough genetic changes will happen to make this flu easily transmissible between humans. We do know that we do not have a vaccine that will work effectively against it and there is no telling if the current vaccines that are in development will be effective against a pandemic version of avian flu because it is still mutating. We also know that the US government has been too slow in preparing for flu and has not stockpiled enough antivirals. At the time this book went to press, the US had around 2.5 million doses of Tamiflu and we are far down on the waiting list to get more from Roche Laboratories because we waited until most other western nations had already placed large orders for it. We still do not have a good plan for distributing it only to those in need. If we only distributed our stockpile to medical professionals treating people with H5N1 in the event of a pandemic, we would not even have enough for them. Not only that, it appears that one of the four main antivirals, amantadine, is no longer effective, probably because Chinese farmers have fed it to chickens for years. In October 2005 scientists isolated a sample of H5N1 from a Vietnamese girl that shows resistance to Tamiflu, the most popular anti-influenza medication. The girl, who had not been around any poultry, apparently got the virus from her brother, who had. Even with this new find, we know we should still be stockpiling antivirals and making new ones more aggressively because that is all we have. Scientists have been saying this for a long time, but it has fallen on deaf government ears *again*.

There are now reports that the CDC has not been releasing much of the genetic sequence data from the thousands of influenza samples that it analyzes each year. This data is extremely valuable to those who study influenza and can provide scientists with the raw data necessary to make significant progress in our understanding of this virus. This is counter to what scientists generally do with genome sequence data, which is to place it in public repositories for analysis. That the CDC has not completely embraced this approach to genome data on influenza virus is baffling.

Run-of-the-mill flu kills around 36,000 Americans each year, mostly the elderly. Yet the government has done little to encourage people at risk to get the flu vaccine every year, has not set up sufficient vaccine production facilities in the US and has failed to invest significantly in the development of new antiviral drugs. Hospitals do not have the surge capacity and quarantine facilities necessary to treat people in the event of a pandemic. The administration is adept at pushing its agenda on many topics, but you never see it doing the same to push flu vaccination or actually getting us prepared for public health emergencies. That's the kind of propaganda I could get behind, because it would save lives, not face. In terms of cost-benefit, the government could set up shop and produce flu vaccine for every American for a few dollars each. That would cost us about a billion dollars, and even if it only saved 20,000 of the 36,000 people who die of flu each year, it would be a huge victory. Our biggest health threat today is not any particular virus or bacterium, but the absolutely incompetent actions of our leaders. This is not up for debate. They dropped the ball. Japan had a plan for dealing with a pandemic flu outbreak as far back as 1997.

In October of 2005, the President held a press conference in which he discussed how he was handling preparations for a potential flu pandemic. In it, he outlined how he might call for a quarantine of a city using the military. Not only would this be impossible in any major American city, it would be completely ineffective. In addition, Bush appointed Stewart Simonson as Assistant Secretary for Public Health Emergency Preparedness at the US Department of Health and Human Services (HHS). Mr Simonson's job prior to coming to the Department of Health and Human Services and running the nation's response to public health emergencies was corporate secretary and counsel for Amtrak. Before that, he was a lawyer for former Secretary Tommy Thompson when he was Governor of Wisconsin. Mr Simonson has no working experience in health, emergency preparedness or really anything that requires the ability to understand medical or scientific information. He was responsible for the Department of Health and Human Services' response to hurricane Katrina, and after resigning from office in May 2006 was given several awards by the Bush administration for his efforts. Nice. Not to worry: the tax cuts that could have gone to fund research and public health preparedness will give you plenty of extra money to buy your loved ones a really swanky funeral if there is a public health emergency.

7

drugs

"...may cause itching, swelling, dizziness, and bankruptcy!"

Americans love their drugs. We take more prescription medication per capita than any other country in the world. As a culture, we treat our medical and personal problems like MTV treats the art of cinematography. We have no stomach for the grit, determination, and endurance required to truly address our personal problems. We opt for pills and elective plastic surgery without making any real commitment to changing the behaviors that negatively impact our health. With such devotion to the quick fix comes consequences, but we are loath to accept personal responsibility for our cultural love affair with medicine. That love of prescription drugs comes with zero understanding of what is required to create new drugs and even less understanding of how and why drugs sometimes don't work or are eventually shown to have serious negative side-effects. So let's consider the basics of drug making.

There are many ways to come up with new drugs, but in the end it is a lot of good old-fashioned elbow grease and luck that determine a blockbuster medication. That's the rub. It's hard and expensive to come up with truly novel drugs. What gets made largely depends upon the available market. Large companies have to produce drugs that are going to make them hundreds of millions of dollars each year to offset the cost of development and satisfy investors' lust for profit. Without the potential for serious cash, drug development becomes too risky. This is because of the extraordinary investment of time and money that goes into bringing a drug to market.

It is estimated that a single novel drug takes 10–12 years and upwards of $800 million to come to market. However, those estimates are based on many drugs that started in the pipeline 20 years ago, when the overheads were significantly lower. Newer estimates put the cost closer to $1.9 billion for a unique drug. Much of the cost comes with the extraordinary failure rate for drug approval. Only about 5 in 5,000 drugs that enter preclinical testing makes it to human trials and only 1 in 5 that are tested in humans is eventually approved by the FDA. That means that there is a 1 in 5,000 chance that a drug that is tested preclinically will ever be prescribed by a doctor. The majority of the cost is actually attributed to the hundreds of thousands of compounds that never make it into a single human.

Makin' drugs

Every drug has a unique story behind it that is a testament to the scientists who believed in a particular compound. Scientists at drug companies do their best to establish methods to find the best compounds to chase, but in the end such efforts are really nothing more than a calculated crapshoot. The hard reality of being a drug developer is that most scientists who work in the drug industry never wind up working on a drug that makes it to market. That alone should be an indication of the inherent risk and cost of making prescription medication.

We now know enough about how different biochemical pathways work in cells that the first step in screening drugs is often identifying a target or a test for changes in metabolic products that indicates that the right pathway is being regulated. It also starts with chemists who design compounds with known properties, like the ability to be absorbed by the body or cross the blood brain barrier. Companies make vast libraries of these chemicals that are often the first step in making a drug. Sometimes the process starts by isolating chemicals from natural sources like plants or fungus, or designing chemicals that look like or are the same as known proteins. It is not uncommon for hundreds of thousands or even millions of compounds to be screened in test tubes, dishes and tiny wells of plastic plates to determine whether they push and pull biochemical or molecular pathways.

Prior to 1962, drug companies only had to show that a drug was safe, but the Kefauver–Harris Bill raised the bar significantly, by requiring companies to submit data on whether they actually work. Once a drug is determined to be effective in cells, drug companies typically move to animal models, and once shown to be effective there, or at least not toxic, they move to humans. Again, there are no hard and fast rules. Along the way there is extensive testing and retesting, but in the end it is in humans where safety and efficacy count. The end goals of clinical trials are to figure out the correct dosage for optimal efficacy with minimal toxicity or side-effects and to quickly rule out drugs that aren't going to make it through.

Before a single person is enrolled in a drug trial the study is carefully designed, including doses to be used, target numbers of people for each phase of the trial, and the ethnicity, sex, behavioral profile, and age of the participants. This is an extremely important process that is designed to eliminate confounding factors that can mask effectiveness and adverse effects. There has been tremendous criticism of the design of trials, but striking a balance is very hard to do. Expanding the numbers in any direction can dramatically increase the cost and risk. There have been some positive signs – inclusion of more women and minorities, for example – but there are still deficiencies in many areas. Those who call for more inclusive drug trials often have little appreciation of the cost of doing a trial and the fact that they are designed to eliminate confounding effects from the diversity of the participants.

After preliminary studies, there are four basic phases of clinical trial that take place using human subjects. Three of them happen before a drug is approved. The first phase involves determining the general safety of the drug. These trials are routinely conducted with 20–100 healthy individuals, often at an in-patient facility so they can be closely monitored. Individuals are generally paid to take part in these trials, sometimes quite well because they last several months and there is a sizable risk because there isn't a lot known about human tolerance. Needless to say, they are tested constantly to see how the drug is being absorbed, metabolized and excreted, so poking, prodding and peeing in cups become as routine as eating and sleeping for participants. The constant monitoring also ensures that if there are any adverse effects the trial can be halted before anyone gets hurt. It will be of little surprise that often college students and poor people tend to sign on to be human guinea pigs in this phase. Between 70 and 80% of drugs make it through.

Taking the knowledge from phase I, which used healthy people, several hundred patients with the target disease or disorder are enrolled in phase II. Some of the participants will be given the drug, while others will be given a control treatment, which can be either a standard treatment or a placebo. It is common for these experiments to be done 'randomized and double blind' meaning that who gets the drug versus who is in a control group is random and neither the researchers nor the participants know who is getting the drug. Safety is still a primary concern, but efficacy is also measured at this point. Researchers will start to get an idea of the ideal dose at which the drug needs to be administered. These trials can last up to two years and only about one third of the drugs that enter phase II pass muster. Many people think that it is cruel to enroll patients and then not give them treatment. But this is essential to measuring efficacy. It's cold, but absolutely necessary.

Phase III is the mack daddy of clinical trial phases. It is the largest (several hundred to several thousand patients with the disease are enrolled) and therefore most expensive phase. Phase III trials are large so that researchers can get a good sense of the efficacy of the drug as well as the range and frequency of adverse reactions that might be expected. Somewhere between 70 and 90% of drugs that enter phase III trials will go on to seek FDA approval.

After phase III is completed and the data are analyzed, a drug company will present the results to the FDA, which will make a decision to approve or reject the drug within 6 months to a year. They can also send the company back to do further clinical trials if they suspect that there may be an adverse effect that might be uncovered through further study or if they want to see how the drug works in a specific population of people.

One of the most difficult aspects of determining whether a new drug is effective during a clinical trial actually has nothing to do with the drug itself. Doctors have known for centuries that a patient's belief that they are being treated for their illness can have a dramatic effect on them getting better. This is pure mind over matter – a positive outlook and the sincere belief that you are going to get better often makes it so. In drug trials, it is standard for some portion of the participants to be given a placebo that has no active ingredient in it so that researchers can tell the difference between the effectiveness of the drug and the 'placebo effect'. The phenomenon has been the dragon slayer to many drugs, but surprisingly little is known about it.

Scientists are now starting to get to the heart of the placebo effect. It has been known for some time that opiods – natural compounds that the body produces that are related to opium, heroin and morphine – play a role in the placebo effect. More recently, it has been demonstrated that neurotransmitters, including serotonin and dopamine, are important in the placebo effect as well. So the psychological effects of being treated can actually push body chemistry and have a direct impact on brain activity, which in turn can have a direct effect on the body.

In recent years, it has been shown that the mere interaction with doctors can cause significant changes that resemble the placebo effect in people being treated with drugs. Since it is important to try to limit the placebo effect as much as possible in clinical trials, interaction with physicians is also limited. Finally, awareness of when and if a drug is being administered is often limited. One study showed that patients who were unaware when a painkiller was administered intravenously showed far less placebo effect than those who were told that a painkiller was being injected.

While it has not been well studied, it is possible that habituating the patient to taking dummy drugs combined with deliberate obfuscation of the nature of the treatment could resolve some placebo effects. If patients are given pills on such a regular basis and told that it is highly unlikely that they are getting the drug the effect could be reduced. This could of course backfire and result in the patients dropping out of the trial, since they are likely participating because they are hoping that they will get the miracle pill that will cure them.

It is also unclear why some people don't seem to show any placebo effect. The obvious possibility is that they have a cold sense of reality or do not believe that they can be helped. But it is also possible that there are genetic or other physiological factors that can have dramatic effects on how people perceive treatment or react to treatment. Finding these factors could tell us a lot about how people process feelings of hope or even despair.

The point of understanding the nature of the placebo effect is at once to harness it in treating disease and to control for it in designing drug trials. Factors like where and how drugs are administered, the type and frequency of clinical follow up, and people enrolling in multiple drug trials influence the placebo effect and can ruin a clinical trial if not controlled.

Money drives the ship

The price and uncertainty of drug development make it impossibly expensive to develop drugs for most diseases. This is because the market for these drugs is simply too small and therefore the potential for profit is nonexistent. So even if a company knew exactly what drug it needed to develop to treat a disease that only affects a couple of thousand people, it would never do it.

Since government-sponsored drug development in the absence of monetary assistance from the pharmaceutical industry is virtually unheard of, rare diseases often do not have drugs developed or marketed to treat them. Enter the Orphan Drug Act of 1983.

Since the cost of getting FDA approval for a drug is the same whether it treats millions of people or hundreds of people, Congress had the foresight to give incentives to drug companies to come up with treatments for rare, so-called orphan, diseases defined in the law as affecting fewer than 200,000 people in the US. The incentives are pretty sweet, including federal grants for conducting clinical trials on drugs intended for rare diseases, tax credits for up to 50% of the costs of conducting clinical trials, and the exclusive right to market the drug for seven years from the date of FDA approval.

That last one is *huge*. Normally, patented drugs are protected for a period of time so that no one can sell the same chemical compound to treat a disease. The exclusive rights clause of the Orphan Drug Act means that no other drug that has a similar effect, no matter what the compound is, can be marketed to treat the disease. This deliberately creates a monopoly on the treatment of individual diseases, but it also means that without competition the price of these drugs could be set at any level, with no chance of competition for at least seven years. There are exceptions (for example if a new drug works better or treats a different spectrum of symptoms for a disease), but on the whole the act crushes competition. Also drug companies are allowed to charge an astronomical amount for these medications, the cost of which gets passed on to the patient's insurance companies and thus indirectly to you. This means that people without health insurance can't afford it and insurance premiums for companies with employees with a rare disease can have their premiums skyrocket.

Take for example, the infamous case of the drug Ceredase, which is used to treat Gaucher's disease. There are only 2000 people in the US with this disease, so it is really quite rare. Genzyme, the company who markets Ceredase, began charging between $100,000 and $400,000 per year for the drug when they brought it to market in the 1990s, depending upon how much a patient was prescribed. They turned it into a huge profit maker and capitalized on the market position that the law provides.

The act has resulted in over a thousand medications receiving orphan drug status, many of which would not have been made if it were not for the subsidies. There is no question that this has been a fantastic benefit to patients with truly rare diseases. But wherever there is a wrinkle in the law, corporations will take advantage of it.

Since the definition of an orphan disease is left deliberately vague, drug companies have been able to deli slice diseases. This means that a drug that is useful against a particular type of cancer, say ovarian cancer, can receive orphan drug status. Later, if the same drug is shown to be effective against another cancer that affects less than 200,000 people a year it can receive orphan drug status for that disease as well. AZT, the widely prescribed AIDS medication, actually received orphan drug status because when it hit the market there were fewer than 200,000 known cases of full-blown AIDS. Once the disease surpassed that mark, the drug did not lose its status because there is no provision in the law to do this. If profits and sales well exceed what is possible for treatment of the disease that it is designated for through new uses or off-label uses, it still retains its orphan status. Several attempts to amend the law have been attempted in Congress, but any attempts to alter the monopoly rules or put price caps on drugs that will still allow profit and the recouping of development expenses have been rejected.

While the Orphan Drug Act has driven companies to develop many effective drugs, and has sparked many biotech companies to be born, it does not serve all diseases. The party line in news articles that describe the discovery of mutant genes that cause rare diseases is that it will 'open the door to the development of new therapies.' This is absolutely disingenuous. Reporters have been asleep at the wheel on this one. Through the lens of reality, journalists would really write 'discovery of the mutant gene for *rarediseasia* opens the door for disease advocates to lobby for more money to do research on the disease with no chance of a drug company taking up their cause.' This is an unfortunate reality of the pharmaceutical industry: for truly rare diseases that affect a few thousand people or fewer, the market is generally dead.

The situation could be corrected by augmenting the Orphan Drug Act to protect consumers and encourage more competition rather than competition-stifling monopolies. There is a longstanding policy that the government stays out of the drug development business. That is because it is widely viewed as taboo for government to compete with business. What if that changed?

In the last 15 years we have made tremendous progress in understanding the basis of many diseases, especially rare genetic diseases caused by mutations in single genes. These discoveries are a direct result of the tools and information gained from the Human Genome Project, the largest public dump of scientific data in history. We are now entering a difficult transitional phase in which information on the cause of disease has to be translated into therapies. A big part of that transition will eventually have to be the cold reality that scientists working outside the pharmaceutical industry will have to get into the business of bridging the gap between basic science research and the development of new drug therapies for disease.

Why not let government scientists and government-funded scientists use tax dollars to develop and produce drugs to treat rare diseases? The cost of the drugs themselves and their development would initially be quite expensive. But in the end, once development costs are recouped through initially higher prices, the cost could be reduced to cover production and distribution. That way, taxpayers and patients would wind up saving a lot of money. More importantly, lives could be saved or dramatically improved. Since these rarest of rare diseases are often simple diseases where replacing a protein in the body would likely suffice for treatment, the normal risk of development would be much cheaper than screening random compounds. Pharmaceutical companies warn that such a precedent would lead to a slippery slope and get the government into making drugs against all sorts of diseases.

OK, let's consider that. Many tropical diseases have little or no treatment available for them and since they do not occur in the US, our drug companies don't aim to treat them. No US market: no drugs. If the US government produced drugs against such disease, millions of lives could potentially be saved and significant trade deals could be made on behalf of US companies based on the supply of cheap medication. Again, the development costs could be recouped through initial prices.

There is one minor catch. For universities and the government, getting into the drug making business is hampered by the fact that they don't know how. There are virtually no researchers who are trained at the graduate and post-doctoral level at universities in how to develop new drugs. There have been tremendous efforts to keep the NIH out of the drug business by drug companies, but we will eventually see a shift.

Most academic researchers have never worked in a drug company and graduate schools do not train scientists how to make drugs. This, however, could change. It is not inconceivable that universities could team up with pharmaceutical companies and set up graduate schools in drug discovery and development. It would be a radical change in the training culture that currently exists in the academic arena, but it would address the hard truth that many scientists eventually wind up working at these companies and that early training could prepare them for work in the industry as well as spark smaller drug companies to form and tackle less profitable, but much needed, drug development for rare diseases.

Students could take classes at the university, but would do their actual lab work at a pharmaceutical company. Universities would have to drop requirements for publications and allow flexibility in thesis reporting to protect drug company trade secrets, but such training grounds could have dramatic effects on the pharmaceutical industry.

Sell, sell, sell

While phase III trials are considered the gold standard, once they are completed and a drug is approved a pharmaceutical company's work is not done. Let's just say that a fictional drug, Lipopaxia, is approved for treatment of a fictional illness: giant ass disease (GAD). The FDA will tell the drug manufacturer what warning labels must be put on the bottle, what dose is approved and for what purpose. The company can now start a marketing campaign and get it out to the public afflicted with GAD and rake in the bucks. But this is not the end of the story.

There is another, more amorphous phase of clinical trials: post-market surveillance, sometimes referred to as phase IV. Here, drug makers will continue to look for adverse effects in the population. They might do more studies to see if it is effective for treatment of other diseases (perhaps giant thigh disease (GTD)), whether the effect of the drug wears off over time, or whether it has an adverse reaction with another drug or food. They might also assess cost-effectiveness against other therapies, though this is somewhat more rare. Finally, the FDA often asks for specific studies to be done on a drug at the time of approval.

In response to public calls for drugs to come on the market faster, the FDA instituted an accelerated drug approval process in 1992 for medications designed to treat life-threatening illnesses. The logic was simple. Drug companies could get a product to market in record time and lives could be saved. There was one major stipulation though: the company would be required to do post-market clinical research trials to further assess safety and efficacy, trials that are normally required before approval. This is an excellent system which has not resulted in any appreciable increase in medications being pulled from the market. However, an analysis in 2005 of these trials found that of the 91 studies ordered on 42 drugs approved through the accelerated approval program between 1993 and October 2004, only 46 had been completed. Even worse, of the studies that had not been completed, half of those had not even been started. Astonishingly, 68% of the drug companies involved with these drugs had not revealed to the Securities and Exchange Commission, that they had been required to perform the studies. This is nothing short of ridiculous. In a time when the public is becoming increasingly skeptical of drug companies' overall motives, when lawsuits are piling up against them, they won't even comply with the regulations that allowed them to bring their product to market earlier than they would normally have been allowed to.

The problem is that the FDA has little recourse against these companies. They have not been able to demand a time frame for completion of the studies or levy massive fines against them for not complying. The business-friendly Congress points the finger at the FDA and is even willing to point the finger at scientists, but has not made a single move to allow stiff punishment of companies who take part in unscrupulous activities.

Post-market survey and marketing of drugs is precisely where our fictional drug company will get into trouble as well. Now that Lipopaxia is approved, they are eager to see it perform well – not in patients, but in the market place. This is where marketing and business people take over and the scientists fade into the background. Their campaign starts off with an army of sales people hitting the streets and approaching hospitals and doctors offices to tell them about the wonders of Lipopaxia. They give out free samples and make sure that a good number of these sales representatives happen to be female and attractive so as not to scare off the fish. In late 2005, the *New York Times* reported that drug companies are actively recruiting attractive college cheerleaders to be drug reps, regardless of their education background. These representatives give out free stuff – pens and pizza to persuade doctors to listen to them, and expensive trips for those who are prescribing it often. Perhaps the trip will include a few seminars on proper use of Lipopaxia and the latest research on its efficacy, subtly implying that it might also be effective in people who have giant stomach disease (GSD) and GTD even though it is not FDA-approved for treating them.

The sales teams know that doctors can prescribe Lipopaxia to treat whatever they want. Just because the drug is only approved for treating GAD does not mean that a doctor cannot legally prescribe it for GTD. In fact, doctors generally have pretty free rein to use their judgment in the prescription drug world. This type of activity is called off-label use and is precisely what we expect to see with drugs like BiDil, which you will recall is only approved for use in African Americans at this point.

The manufacturer of Lipopaxia is increasingly aware of the power of television and therefore also starts an aggressive television ad campaign to inform the public about the drug. Since there are tight regulations on these ads they just tell you that it is approved for GAD and show some happy slim people frolicking on the beach as they tell you that you should ask your doctor about Lipopaxia and about living a full active life with GAD. The rapid fire list of possible side-effects and exceedingly fine print that flashes quicker than the average person can read does not obscure the message: if you are suffering from anything resembling GAD, GTD or GSD, you should ask your doctor about the new wonder drug. You might even be forceful with your doctor about your desire to try Lipopaxia. This is your ticket to normal-sized underwear (the anal discharge side-effect is a small price to pay).

Your doctor tells you that you don't actually have GAD. He insists that you are just overweight and only have some of the symptoms of GAD, but are lacking the metabolic disorder that causes it. He suggests an exercise program and a reduced calorie diet. This is frustrating to you because you have already explained that you have tried and failed. You press harder to get the pills, but he is resisting because he knows that it can cause irregular heartbeat, dry mouth and of course anal discharge. Finally, he gives in and prescribes the Lipopaxia because getting thin in your case is probably going to be better for you in the end. After a few months you go back for a blood test or two to make sure there are no complications (and to get a new prescription). Your doctor sees that you have lost a remarkable amount of weight and are now far more active. Your hypertension seems to have dissipated somewhat, so he gladly refills the prescription. After all, you do appear to be more healthy. Slowly but surely, he starts prescribing it to many of his overweight patients because he has heard similar results from other doctors, and the Lipopaxia sales woman (who happens to be totally hot) is now giving him little extras for prescribing the product and telling other doctors about his patients' success.

Two years later, reports of people taking Lipopaxia developing liver disease start cropping up. There aren't that many cases (maybe 1 in 800 patients), but it is worrisome despite the fact that the incidence of liver disease is only slightly lower (around 1 in 900). Since the people taking the drug tend to be obese, and thus might have underlying diabetes, which is a risk factor for liver disease along with alcoholism and hepatitis C infection, there might not be cause for concern. Some of them have had to be put on the list for liver transplants and some have died from it. Sure enough, the sporadic reports start piling up and scientists at the company that makes Lipopaxia start to worry. More reports come in and it appears that the link could be real in some patients. Internal memos talk about a possible link between the two, but nothing is concrete yet, even though the incidence is now certainly higher than that of the general population. They do the math and figure that the increased lifetime risk for liver disease only goes up by 5% among people on this drug. They continue to look into it and as it turns out, this is real – rare, but real. People taking Lipopaxia are more likely to develop liver disease than people in the general public. It isn't yet clear if these people are diabetics, if this increases the risk in just some people (say alcoholics) or if this is an anomaly. There are some within the company that are very worried that if word gets out it could hurt sales. They press for more data, which takes time to gather and analyze. In the meantime, they keep quiet.

The sales woman continues to sell the drug at breakneck speed. She is even given bonuses for selling so much of the stuff. Your doctor continues to prescribe it because his patients are doing fine and the link between the drug and liver disease is still not in the public arena and he has only heard rumors of the link. No published scientific evidence points to any unexpected danger. In his mind the drug is still making people healthier by reducing their weight. You have switched to taking it less frequently to maintain your new figure and feel much better than you did before. This too is off-label use, but is working great. Your doctor is OK with it because lots of people are doing the same thing. This seems to be a miracle pill.

Over a period of months you start to feel more tired, which gets worse with time. Slowly you start to get itchy skin and develop spider-like veins, but worst of all your urine is getting really dark and your stool is getting a pale color. Finally you go to the doctor who does some blood tests, which show you have developed liver disease. He immediately checks you into a hospital and takes you off Lipopaxia. In a few weeks you recover, but not without a huge hospital bill and possibly permanent damage to your liver and kidneys.

A few months after you are released from the hospital, the newspapers start reporting that Lipopaxia might be causing liver disease in some people who have been taking it for several years, and that the drug company knew about it but did not tell the public or pull the drug from the market. But they downplay these cases and say that they do not believe that most people taking the drug are in danger. Your blood boils and your immediate response is to call a lawyer.

This is precisely the type of scenario that has been played out in courts over the years. But whose fault is it? The pharmaceutical company? I think not. While their sales techniques are to be questioned, it was your politicians that allowed them to get away with it. That aside, the blame falls right on the shoulders that hover above your fat ass. You shouldn't have been taking the drug and you knew it. Your doctor warned you and instead of making the lifestyle changes to slim down, you took the easy route. Don't worry: your doctor is also a jackass. He knew you didn't have the disease and knew that you didn't need the drug to get thin and could have stuck to his guns and said 'No.'

∿∿∿

There are three basic types of drug: those that are isolated from living material, those that are chemically synthesized to be identical to naturally occurring compounds or proteins, and finally those that are chemically synthesized, but are not identical or necessarily similar to naturally occurring compounds. Traditionally, drug manufacturers have focused on the last category. Advances in small molecule chemistry in the 1970s and 1980s had a significant effect on the industry and resulted in a tremendous increase in new drugs coming to market in the late 1980s and 1990s. Drugs that are identical to or are specifically designed to mimic the shape of proteins are the latest innovation in drug design. This is a direct result of the information that poured out of the Human Genome Project and the explosion of molecular data that has come out since the early 1990s on the function of proteins and genes.

There has been a gradual decline in the number of drugs in clinical trials and seeking FDA approval over the last decade. Analysts are attributing this to two problems: corporate mergers in drug companies that are more and more focused on making billion dollar a year drugs, and the natural shift to protein-based drugs and drugs that are based on specific information about molecular and genetic factors associated with disease. So the shift has resulted in a natural slowing of the pipeline as drug companies learn how best to utilize the information available. This is a significant advance over the massive screening that characterized much of the previous drug isolation process. One of the best examples is insulin and human growth hormone, which used to be isolated from animals. Now these two proteins can be made in cells, which is both less expensive and far safer.

Most of the drugs that have come out from these efforts can be considered 'low-hanging fruit' in the drug world. Many of them were proteins known to have a specific function that could simply be replaced or supplemented. This class of drugs is often referred to as biologics. Most of them are simply copies of proteins that normally exist in the body. Biologics are distinct from chemical drugs because they are isolated from living material. This class of drugs includes vaccines, antibodies and other proteins generated through recombinant DNA technologies. As genetics and genomic medicine progresses, you will see a lot more of these drugs on the market.

Lately the issue of generic biologics has come to a head because it is estimated that $10 billion in branded biologic pharmaceuticals went off-patent by the end of 2006. This created opportunities for cheaper generic versions of these compounds, which are very expensive to produce to come to market. Since approximately one half of all prescriptions filled last year were for generic drugs, but they only accounted for 8% of money spent on prescription medication, there is a real opportunity for saving consumers money if biologics were opened to generic production.

The main issue here is the need to prove therapeutic equivalence. In essence, this means that you have to prove in clinical trials that your drug is as effective as the original product. Generic chemical drugs do not have to undertake these studies. The pharmaceutical industry has argued that it would be harder to duplicate the active compounds in biologics than standard chemical drugs and therefore you cannot guarantee the safety of generic biologics unless you do these types of study. It is true that it is harder to make these compounds, but untrue that you could not reproduce the active compound and ensure through standard diagnostics that it is the same as the original without clinical trials.

The real issue comes when reproducing the carrier, filler and coating compounds that go along with drugs. These compounds are trade secrets not covered in the patents for the active ingredient and they are essential to the performance of all pharmaceuticals. Simply put, they regulate absorption of the drug into the system. Many biologics require a narrow window of concentration in the bloodstream to be effective and therefore changes in absorption rate could make the biologic compound toxic or useless. This problem is not unique to biologics. It is a major ongoing issue with all generic pharmaceuticals. However, it is certainly a bigger issue for biologics because they are quite finicky about how and when they are absorbed.

Biologic drugs should probably be considered for generic approval on a case by case basis, depending upon the complexity of the drug and the active ingredient. In many cases the market and the potential profit are large enough to drive generic development including the burden of providing evidence of biologic equivalence. Generic producers would still have a lot lower development cost than the original manufacturer even if these studies are required. Diseases where a change in drug efficacy (i.e. absorption) could have very serious acute side-effects will absolutely need therapeutic equivalency studies. That need could be determined by the FDA through a regulated approval process. Therapeutic equivalence studies would be prudent for generic biologics designed to treat thyroid diseases, anemias and epilepsy, for example, because they have a very small window of efficacy and could be quite toxic at too high a blood plasma concentration. The FDA and Congress should definitely lean towards caution in laying out the regulations for the production of generic biologic drugs, but have been slow to put out definitive guidelines because of the wide range of issues including the possibility of requiring drug makers to reveal everything about a particular medication, including coatings, fillers and carriers.

Good drugs, bad drugs

While the details vary from case to case, the point is that pharmaceutical companies and the FDA have been bearing almost all of the responsibility at the behest of finger pointers. This is wholly unfair. It is absolutely ludicrous for anyone to think that taking prescription medications is an OK thing to do. It is not a reasonable alternative to changing behavior or for taking over-the-counter medications, which are on the whole much safer. Doctors are to blame as well. Many of them have become way too comfortable doling out prescriptions to people who are not sick. Whether it be antibiotics, weight loss pills, anti-anxiety pills to unhappy homemakers or sleeping pills, too many physicians are turning into Dr Feelgood.

Besides trust issues between medical professionals and patients we have an underlying trust in the safety and effectiveness of medications. There has been tremendous media attention (and thus political attention) paid to a handful of high-profile cases in which prescription medications have been linked to severe adverse side-effects. This has given the public the impression that drug companies and the FDA are not doing enough to protect us. It is rather amusing (and perhaps sad) that so few of us remember the public outcry that it was taking too long for new drugs to be approved by the FDA and that the pharmaceutical companies were having to jump through too many hoops to bring drugs to market. Granted, many of those outcries came at the behest of drug companies, which complained about having to conduct expensive trials to ensure the efficacy and safety of their wares. But there was significant posturing on the part of patient advocates and politicians as well. Here we are, but a decade later, and we hear a very different tune.

Like the ubiquity of mistakes in the practice of medicine, unexpected adverse side-effects (some serious, some not) are common in pharmaceuticals. But Americans have bought the idea that when the FDA approves a drug it is guaranteed to be safe. This is certainly not true. We need to accept that the onus is on drug developers to make a product that is effective and safe within the confines of limited clinical trials. However, some negative side-effects only appear over a long period of time in certain populations or in combination with specific environmental factors or as an adverse interaction with other drugs. In lay terms? Shit happens. Deal with it.

This will come as little comfort and perhaps sound a bit callous to those who have suffered from adverse drug reactions or who have even lost a loved one due to a medical mishap. These words will be especially painful to those who were injured or who became ill through no fault of their own or who had been following what they thought was the best medical advice available. The idea is meant to convey the 'certain uncertainty' that goes hand in hand with medicine and drugs. The question that must be addressed is how to reduce such incidents before they happen, while still allowing new medication to come to market.

This brings us to Vioxx. This is a good example of how far the drug industry has come and how crazy people have become in response to adverse drug reactions. I'm actually still a fan of this drug, as well as of other painkillers that act as so-called cox-2 inhibitors, despite the fact that they were pulled from the market. Think I'm crazy? Let's go for a walk.

There are reports that Merck executives hid the fact that they knew that their drug caused a significant increase in the risk of heart attack and stroke for some time and even fought off the FDA scientists who originally wanted Merck to put a label on the drug that warned of the risks. It has also been reported that all the way to the top of the chain they were putting undue pressure on, and even threatening scientists (who used to work for them as consultants) because they spoke publicly of the risks associated with the drug. Some of these reports tell of calls to the deans of medical schools, making what have been interpreted as threats to pull funding for research. The reports also say that Merck aggressively marketed the drug for off-label use to doctors and the public in various forms to increase the bottom line. But what about the drug itself?

The individual risk of having a heart attack or stroke on any given day is very low. The drug seems to increase that rate by a factor of 3 and is estimated to have played a role in approximately 30,000 heart attacks, strokes and deaths. But consider this: there are a subset of people who have severe adverse reactions to other painkillers. Their stomach and intestinal tract just can't handle them. For people suffering from chronic pain, the choice is either to live with it or to go on opiates, which are quite addictive. Vioxx, Celebrex and Bextra were life-changing drugs for these people. For them, a three- or four-fold increase in the risk of heart attack is nothing compared to the alternatives. Not only that, it appears as though the risk is associated with long-term use of the drug. So taking it for a few days or weeks in people with no indications of cardiac problems might not actually cause any harm.

I feel terrible for the 30,000 people who died after taking this drug and their families, but I somehow doubt that the events that surround Vioxx will serve as a wake-up call to the public about the inherent risks associated with taking any prescription medication. You will continue to see a lot of smoke and mirrors on Capitol Hill, in the FDA and even at drug companies about reforming their behavior, but I think it should be made clear that the people who spend their lives making drugs for a living never make a piece of the massive profits that companies make off their inventions. They aim to do good. What we need is serious reform of how Americans view drugs and how the industry is allowed to market its products.

I am sickened by the doctors who rely on marketing seminars and visits from cute sales reps with free samples to make decisions about what to prescribe their patients. I am also sickened by a culture and political atmosphere that allows the marketing of prescription medication to the public. 'Ask your doctor' and a list of side effects is not acceptable when we all know damn well that these sentiments mean nothing to a public that is looking for a quick fix to their problems. Finally, I am sickened by juries who are unable to decipher the medical facts in drug lawsuits and try to teach companies a lesson by awarding multi-million dollar settlements.

The hypocrisy in our national drug policy never ceases to astound me. Take for example the medical marijuana issue. Access to medical marijuana is a minor healthcare issue that has been blown out of proportion because of the often illogical stigma associated with the illegal use of the drug. A large part of the problem lies with the extremists on both sides of the issue, who continually stick their finger in the chest of the other side. First there is the pro-medical marijuana lobby, who have been accused of using it to get a foot in the door on the way to legalizing marijuana for recreational use. In truth there is a large part of the marijuana lobby that really does feel this way and which really is using the issue as a battering ram on the door of the larger legalization issue. On the other side are conservatives who see legalization of medical marijuana as a slippery slope to reefer madness. This is reflected in the Bush administration's policy, enacted through the DEA, which considers marijuana to be a gateway drug to addiction and crime. The DEA now focuses a disproportionate amount of energy keeping marijuana off the streets compared with other drugs directly implicated in violent crime and addiction.

In 1994, a close friend of mine was diagnosed with throat cancer. Since he was quite young and in good physical shape (outside of having cancer), his doctors at Sloan Kettering Cancer Center in New York enrolled him in a very aggressive treatment schedule of chemotherapy and radiation. I had just entered graduate school and could only see him on weekends because of my heavy workload, but I looked forward to cheering him up. Befitting our similar dark sense of humor I prepared for my first visit to the hospital by collecting biohazard and radiation signs from the lab with the intention of decorating his bed and IV stand to indicate his new status as a living radiation/biohazardous waste spill. He had gone through a few treatments already, but this one had really thrown him. Besides the pain and discomfort associated with pumping poison into your body, which is essentially what chemotherapy is, he was having a very difficult time keeping down food. This had resulted in rapid weight loss, which is common, but dangerous in cancer patients. So, along with my bag of assorted gags, I brought some marijuana. My friend had been given a variety of drugs for nausea, including one that is a synthetic version of the active component in marijuana (THC). None of them had worked.

The synthetic marijuana derivative was actually a pill, which he promptly vomited up.

The United States Drug Enforcement Administration lists marijuana as a Schedule I drug, which means that it has high potential for abuse and has no legitimate medical use. The assignment of marijuana, but not tobacco, as a Schedule I drug seems completely arbitrary. Nicotine as administered in tobacco has a high potential for abuse, is one of the most addictive substances known and has no legitimate medical use. Nicotine causes a sense of euphoria that, while different in mechanism from marijuana, is far more dangerous because of its addictive qualities. Cigarettes are nothing more than a legal drug that kills people, and we are completely hypocritical in our approach to controlling it.

When I got into New York, I noticed a couple of patients in wheelchairs around the corner from the entrance to the hospital smoking. As I got closer it became apparent from their behavior and the smell in the air that they were smoking pot. Cancer treatment hospitals have long recognized the fact that marijuana helps control nausea associated with chemotherapy and can improve the appetite of patients, but since marijuana is illegal they cannot accommodate its use. So cancer patients are forced to hide their use of marijuana like naughty kids. There are a lot of anecdotal reports of hospital staff and even police turning a blind eye to marijuana smoking amongst patients and that some medical staff will actually suggest it for patients having a particularly difficult time. Hospital policy is very clear on this, but it is hard to deny the evidence that medical professionals will, in this case, put compassion ahead of hypocritical laws.

Another friend of mine, who is HIV positive, travels with a small pipe and a bag of pot at all times because the drugs he has to take to keep his viral load down can cause nausea and reduce his appetite. Weight loss is common in HIV patients, and, as in cancer patients, it can be very dangerous. Many HIV-positive people will tell you that marijuana is the most effective drug for increasing their appetite and reducing nausea. In my friend's case, he will smoke a tiny amount in the morning to increase his appetite enough so that he can eat during the day. He is extraordinarily careful not to smoke too much because he does not want to show up to work high. His employer knows of his HIV status and marijuana use and is understanding and supportive. But this is not always the case. There is still significant misunderstanding of HIV in the US. But one thing that is now very clear is that people can live for decades with HIV due to the enormous success of antiviral medications. Therefore drugs that help patients maintain an appetite are crucial.

Both sides have legitimate arguments. It is true that marijuana use is prevalent amongst those who abuse harder drugs like methamphetamine, cocaine and heroin. But that doesn't make it a gateway drug. Any legitimate drug counselor will tell you that the overwhelming majority of people who smoke marijuana do not move on to harder drugs, but people who have other significant angst in their life will very often abuse hard drugs and start using drugs at a younger age. At the same time, marijuana is the least addictive drug on the DEA Schedule I list and its use is not tightly linked to violent crime (unless you consider devouring a box of Twinkies a felony). But these arguments completely miss the point of using marijuana in medicine.

There is no question that marijuana is effective in many patients for reducing nausea and increasing appetite. Arguments that it will lead to a spread of abuse by people getting hold of prescription marijuana are pretty silly. If it is classified as a Schedule II drug, the same category as cocaine and heroin, it can be very tightly regulated. Allowing patients to grow their own will also not result in rampant use if it is regulated, and if those with permission to grow it to treat disease are registered.

The American Medical Association is strongly encouraging research into medical marijuana to determine the proper dose and regulation of the drug, but there are few doctors that would challenge its efficacy. The only problem is that scientists trying to do clinical research with marijuana are being denied access to it. They are given federal grant money to do the work and then denied the actual drug for the research. Questions about dosage regulation become largely moot in the face of someone not being able to tolerate food or pills. Self-regulation would, of course, lead to people smoking it to experience euphoria. Personally, I couldn't care less if a cancer patient decides that they want to feel euphoric while going through chemotherapy. One only needs to see the anguish on their face once to realize that reform is needed.

Drug companies are now developing THC inhalers and patches, which if effective and approved by the FDA would at the very least dissolve the medical marijuana issue, but until such delivery methods are established it is absolutely unconscionable and sort of pathetic that we use rhetoric and fear to deny people a legal method of obtaining it. In the meantime, we are forcing HIV-positive and cancer patients to find illegal methods to secure marijuana. This is the tip of the iceberg, but a shining example of how dumb we have become.

A natural conspiracy

You've been to your doctor over and over again, and you feel great frustration because your GAD has only gotten worse and it seems to be spreading. You suspect that you might be developing GTD, but cannot be sure. You feel desperate and will try anything to cure it. Since Lipopaxia was taken off the market you have felt rather hopeless. Dreams of being able to fit in an airplane seat without the seat belt extender have slowly faded. One night, you are up late polishing off a pint of Ben and Jerry's and taking occasional breaks to reach for the remote. Thinking you have found some soft core porn on one of the movie channels, you flip back and discover an infomercial for a dietary supplement. In it, the company claims that you can lose weight by increasing your metabolism, that you will not want to eat as much and will feel great, all without the pain of exercise and dieting. All you need to do is call. Since they clearly said that their supplement was totally natural, you assume that it is safe. In a moment of weakness you make the call and give them your credit card number.

In 1994, the US Congress passed the Dietary Supplemental Food Act, which deregulated the dietary supplement industry. Since then there has been an avalanche of new dietary supplements introduced to the market. The companies that market these supplements are allowed to say almost whatever they want about the supplement as long as they do not claim that it cures or treats a specific disease.

Undeterred, supplement companies started to use creative language to sell their snake oil. Take vitamin A, for example. Companies cannot claim that vitamin A helps restore vision loss or that it will enhance vision, so they simply state that vitamin A is 'essential for proper vision'. Consumers do not read into the nuanced language – they just buy it and expect results. Not only are the implied claims of effectiveness disingenuous; the actual content of supplements is completely unregulated because the FDA has no power to pull a product if it thinks the content claims are bogus – that is the job of the Federal Trade Commission. But the FTC is not concerned with the content of the pills, just the labeling and advertising.

A recent Harris poll showed that most Americans believe that if a dietary supplement is sold, it must have been approved by some government agency, that a company can only make claims if it provides clinical data in support of them and that companies are required to warn potential customers of potential health risks or side-effects of their product. As it turns out, none of this is true. The dietary supplement industry lobbied Washington very hard to make sure that the

Dietary and Supplemental Food Act guaranteed that the government had virtually no right to regulate their actual products. The FDA can only act on dietary supplements if there is strong evidence that it is harmful to health. So, unlike all prescription medication, the onus is on the FDA to prove that a supplement is dangerous. They have no oversight of the content of the supplement, consistency of the product or effectiveness. So even if an overwhelming amount of data is presented that the ingredients of a supplement provide no benefit to health, they cannot take any action.

The FDA and scientists who actually understand that supplements can be dangerous to health have expressed great frustration over the situation. To maximize the little protection they *can* provide the public, the FDA has chosen to look at reports on the safety of specific ingredients rather than specific products. This is because a company can just withdraw a product and release it under another name or via a subsidiary if the company is targeted for investigation. Since companies do not have to report any adverse effects associated with their product, the FDA has to rely on publicly funded research into the effectiveness of and possible dangers associated with different products. In essence, tax dollars are doing the work that companies traditionally have to do when bringing a consumer product to market.

When your package of supplements arrives you quickly open it up and are on your way to the normal sized underwear you had when you were on Lipopaxia. The bill for the pills is enough that you briefly think about how you are going to afford all those new clothes you will need. A quick calculation on cost-cutting on your kids college savings plan and you're all set. You quickly discover that the bottle has no information on possible side-effects or the amount of the active ingredient, but you follow the directions anyway and take one pill before dinner. You follow the routine, continue taking the pills and start to feel better. One side-effect is that you find it hard to sleep some nights, but that is a small price to pay and you can get a prescription for a sleep aid from your doctor anyway.

Three months into the regimen and you have lost a total of 5 pounds and are starting to suspect that this pill is no miracle. This moment of clarity is your first indication that you have been swindled. You go to the Internet and find that four of the ingredients are considered stimulants. The truth is you bought a pill with caffeine in it, which gave you more energy, which motivated you to exercise, but it did not melt the fat and did not change the molecular structure of your cells. It did boost your metabolism, but only during the time the caffeine was in your system. The cold sense of self-realization washes over you when you realize that a cup of coffee would have had the same effect.

The FTC has had its hands completely full chasing the false claims that dietary supplement companies make about their products. In 2002, it found that more than half of all dietary supplements had exaggerated or false claims made about them in their advertising. But the supplement industry is savvy and is now far more careful to imply effectiveness without stepping over the arbitrary legal line. This does not mean that it isn't still ripping people off though.

Not long ago I went into my local 'nutrition' store on a bullshit hunt. The door had not even closed before I realized that almost every bottle and advertisement had manipulative language designed to stretch the truth. For example: a bottle of apple cider vinegar (natural source of vitamins and minerals important for maintaining health), vitamin A (essential for normal vision and necessary for eye health), Ginkgo biloba (helps support blood flow to the brain and supports mental sharpness), and echinacea (helps support the body's natural resistance). Other products were essential for liver function, detoxified the liver, oxygenated the blood, and enlarged the size of cells. Each of these products walks the legal line, but in reality the statements made are more akin to fraud than reliable health information because they are designed to imply that they will do something that they have not been shown to do. Is vitamin A essential for normal vision? Yes, but will taking it improve your vision? No. Does Ginkgo blioba improve memory and mental acuity as has been touted? Nope. Is apple cider vinegar going to provide you with the nutrients you need? Maybe. But so will an apple, for much less money than the stuff in the store.

Echinacea is one of the worst culprits. This stuff has been touted as a cure for the common cold, or at the very least has been claimed to be effective in reducing the intensity and duration of a cold. Study after study has shown that it is ineffective against colds and often contains less ingredient than is advertised on the bottle. Nonetheless, literature in the store directed people to take it every two hours at the first signs of a cold to reduce the severity of the illness. The woman in the store pointed me right to it when I asked what to take for a cold and told me 'this stuff works great.'

Many supplement companies have started to realize that there are some people who want to hear that their product has been clinically tested for efficacy. They are less concerned about safety than about being ripped off. So, many companies have started to make such claims about their drugs. (I am starting to get tired of calling them supplements when they are actually drugs.) Few of them will tell you what they mean, but when they do reveal their evidence, it is rarely anything that even approaches a legitimate clinical test. Some of them have even gone so far as to conduct their own 'clinical trials'. I could not find a single one that had ever been published in a peer-reviewed journal. This is not surprising, since they are completely laughable. One I found had followed eight people taking a weight loss pill. Others openly, or not so openly, extrapolate efficiency from legitimate research, making exaggerated claims that the scientists who did the research would never have made. These companies can take comfort in the fact that even if they were forced to post the results of their 'research' (which they claim supports the efficacy and safety of their product) on the web, the public is too stupid to know how to judge it. Frankly, even if the public had a sound educational background that allowed them to comprehend basic science it wouldn't stop them from buying the product because, more than anything else, companies that market nutritional supplements are selling hope, not cures. This is nothing more than parasitic fraud.

Ripping off the public is not the only danger in this industry. Some of the products have been shown to be downright dangerous. In 2004, the FDA was finally able to ban the inclusion of ephedra in dietary supplements. Pills containing ephedra were touted as weight loss miracles and energy boosters. It was the most popular of supplements because it gave people a rush of adrenaline that made them feel energized. It accounted for about 10% of all dietary supplement sales, totaling over a billion dollars. There was this tiny issue though. Ephedra use, especially in packages that are not required to standardize the dose, had been linked to heart attacks, strokes, anxiety and a host of other nasty side-effects. The fact that it is isolated from a Chinese herb called *ma huang* did not make it any safer.

This is the point: these products are claiming to contain and often do contain pharmacologically active compounds. Treating pharmacologically active products called supplements differently from pharmacologically active compounds called drugs makes no sense at all, especially because they often are pushing and pulling the same biochemical pathways. Ephedra is especially weird because other ephedra-like compounds found in cold medication have been tightly regulated since the 1980s. On top of all that, it was well known at the time that the Dietary Supplemental Food Act was passed that ephedra could be used to make methamphetamine, one of the most dangerous abused drugs known because it is so terribly addictive. So much for the war on drugs: let's make it easy to get lots of the key ingredient. Limiting the ability of the FDA to regulate it was directly responsible for the deaths of thousands of Americans. Put in perspective: more people died of ephedra-caused heart complications in any given week than died from anthrax in 2001.

Issues of safety are not just related to adverse side-effects from the compound itself. Take St John's wort, for example: this unfortunately named supplement is basically ground-up plant. It has been touted as a 'cure' or 'treatment' for depression by many in the industry, and there are thousands of web sites dedicated to it. There is conflicting evidence that St John's wort has an effect on depression, so effectiveness is an issue, but more disturbing than that is the fact that the likely active ingredient, hyperforin, acts as a serotonin reuptake inhibitor. In English, this means that it works on the same brain chemistry as Prozac, Paxil, Zoloft and a host of other antidepressants. It also happens to affect the body's ability to process approximately 40% of all prescription medication by interfering with the regulation of the genes that control their normal processing.

Finally, there have been several research studies that have shown up to 100-fold differences in the amount of the active ingredient in St John's wort and wildly varying consistency from batch to batch from the same brand. All this adds up to a very dangerous situation, where people taking prescription medication and St John's wort can have a very bad drug interaction and can experience wild mood swings when going from one brand to the next or when opening a new bottle.

All attempts in Congress to put tighter restrictions on supplemental foods have been thwarted by Senator Orrin Hatch of Utah, the lead on the Dietary Supplemental Food Act that deregulated the industry. Not coincidentally, many of the largest supplemental food companies happen to operate in his home state, and the Senator's son is reported to have close ties to the industry as well. The Senator now supports reporting of adverse side-effects from nutritional supplements to the FDA, but has done little to ensure that the products are safe in the first place.

As culpable as the Congress and supplement companies are, the modern day snake oil salesman is the industry's pimp. There are now thousands of books, infomercials, web sites and print ads oozing over the information landscape making fraudulent claims about how nutritional supplements will cure disease and heal the sick. One character in particular is a used car salesman with no formal medical training. He was brought up on fraud charges by the FTC and on credit card fraud and served time! He claimed that there was a machine that could be dialed to a specific frequency that would rid the body of cancer. His books have sold millions of copies.

It may sound like I am demonizing all nutritional supplements. I am not. I am demonizing the rampant hyperbole within the industry. Many products put out by reputable companies are quite useful and do improve health. This is mainly because Americans have such shitty diets and need to go to pills to get the nutrients they need (I include myself in this). I am saying that the industry has been unwilling to police itself and Congress is 100% culpable in the proliferation of the massive hype the industry spews to the citizenry, endangering their lives and ripping them off.

8

healthcare

Our current medical system is a dysfunctional, adversarial, co-dependent scrum of insurance companies, practitioners, patients and the government. At the heart of the scrum is money. There are three basic issues: (1) waste, in the form of fraud, and unnecessary treatment; (2) insufficient preventative or appropriate care, which leads to large increases in healthcare cost when complications or disease set in; and (3) insurance, which can limit care. Each of these areas directly impacts the others, causing a cycle of increased cost that is passed on to consumers. However, the most difficult problem to resolve has been insurance. It is the hub of all medical costs in the US and is also the lynchpin of the political fight, where different philosophies clash: one pro-business, the other pro-access and care.

It is easy to get lost in the maze of regulations and laws that affect the insurance game, but it all boils down to ideas of social insurance and limited insurance. Most western countries, including the US, have been based on a social insurance policy where everyone pays about the same for healthcare, regardless of their health status. For example, a twenty-something health nut who hasn't seen a doctor in years pays the same as a fifty-something suffering from heart disease, diabetes and gout who sees the doctor on a regular basis and takes a library of prescription medicines. The system is set up to ensure that healthier people in essence wind up paying for the insurance of sick people, but are assured that if their health is compromised, they will have access to healthcare. This system of social distribution of cost works well everywhere but here, because it has slowly been infected by another philosophy where your insurance rate depends upon your use of the system.

This is in fact the second pro-business philosophy that has taken hold of the healthcare industry through changes in insurance policy. The root of it is a concept called 'moral hazard', which states that access to healthcare unnecessarily increases the use of healthcare; i.e. if it's free and cheap then people will overdo it. The principle is overly simplistic though. It presumes that people like to go to the doctor, which we all know is untrue. Rich people don't go to the doctor more than people with normal health insurance and people in government-subsidized healthcare plans like Medicaid and Medicare do not use the healthcare system more than people of equal health who have the best insurance money can buy – plain and simple.

Nonetheless, insurance companies, which continue to make record profits, have been freely reducing access to healthcare procedures and medication, which increases their bottom line. They then increase the rates for employers whose workers incur high healthcare costs due to severe illness, thus treating healthcare in much the way that we treat car insurance. If you are a bad driver, you pay higher insurance than people who are good drivers. But healthcare is not the same as driving cars. Sure, most people who are in bad health are so because of their own behavior, but not everyone. The system punishes bad health by raising the cost of health insurance for these people. Here is where the feedback starts: if you limit care or raise costs, people are less likely to go to the doctor when they feel ill or 'choose' not to have insurance. It is absolutely accepted in the medical field that the most expensive procedures, including hospital stays, increase in frequency as a function of neglect of health, especially through not getting adequate care early on.

The 'moral hazard' philosophy is, not surprisingly, the center of healthcare reform today. President Bush is pushing his version of moral hazard in the form of Health Savings Accounts, where people pay for their own basic healthcare through tax-free accounts. These accounts are favored by wealthy people who have disposable cash to bank for health care expenditures. They obtain coverage for serious health matters through private insurance plans that have big tax advantages and high deductibles. So people in relatively good health will choose cheaper, stripped down plans that offer fewer services, and people in bad health will be forced to pay more for their care. This is a dangerous place to be, because if a healthy person gets severely sick and needs services outside of their insurance coverage, it comes out of their own pocket. It completely ignores the fact that the most common reason for bankruptcy in the US is health-related bills, and will most certainly add to the problem.

So, what is driving this trend? Some say it is flat-out greed, but there are those who legitimately feel that this is the most sensible way to reduce healthcare costs. They even roll out studies to prove it, but these economic studies are nothing more than bad logic, false presumption and a fundamental lack of understanding of why people seek medical care. At the center is the so-called 'evidence' that moral hazard applies to healthcare. For example, the Bush administration frequently cites the 2004 Economic Report of the President when talking about healthcare. The report claims that the reason for the rise in healthcare cost is over-consumption, and details the fact that people really don't want health insurance (no, I am not kidding). It claims that one fourth of the people who do not have health insurance had it available to them through an employer and chose not to be covered. It also claims that many of the uninsured were young healthy people who just didn't want coverage. They also talk about people being eligible for Medicaid but not applying, and about others being foreigners working here who have coverage outside of the US.

Extrapolating, the report is implying that a high percentage of the 45 million people without health insurance just don't want it. It fails to recognize that the insurance that these people are turning down is often too expensive for them, and that given the choice between paying their electricity and food bills and having health insurance, these people eat and have electricicity. Fifty-six per cent of the people who don't have health insurance are families with at least one full-time worker. Only 19% of the uninsured are not in families where no one is working. So the report is right in one sense – they did 'choose' not to enroll in available insurance – but they are doing so for the same reason I don't own a Ferrari. Never mind the fact that half of people without health insurance owe money to hospitals and that fifteen million Americans without health insurance are now being chased by collection agencies.

'George Bush doesn't care about black people'

For a minute, let's try to think about the complexities of healthcare as if we inherited our money from our daddy and never had to fill out a health insurance form or randomly pick a doctor from the rolls of an HMO.

In Bush logic: the problem is with people, see. Not all people, just good people who are confused. People without insurance go to the doctor less. They spend about $930 a year on healthcare. That's efficiency. They use what they need because they can. Spoiled folks who have these luxurious health insurance plans spend about $2350 a year on healthcare. By figur'n a bit that means that people with insurance are spendin' about $1400 a year on wasted healthcare. You see, people are scared of their health, so they go to docs more. That's waste. If we make 'em pay for their waste, they'll stop waste'n .... These people are also gonna sue their docs. We gotta stop that too so they can get on with their important work. Where's my chain saw?

Fact: the public thinks that every American citizen should have health insurance and sees repairing our healthcare system as a top priority. There have been six serious attempts in the government to establish universal coverage of Americans, and each time it has been fought off by conservatives. The most recent attempt was during the Clinton administration where then First Lady Hillary Clinton tried to engage Congress in serious discussions and began formulating a plan. That plan was squashed by Republicans in Congress, who pointed to the so-called failures of the Canadian system, using scare tactics to crush the idea and calling her a failure. After all, we wouldn't want to have a plan like the Canadians, Germans and Japanese have. For Christ's sake, Canada is a frozen wasteland and its citizens are mauled by polar bears in the streets of Quebec every day. Do you want polar bears roaming the streets of your town? I exaggerate, but they really did use a smear campaign and rhetoric rather than rational discourse on healthcare. The Republican Congress did not take a single step to increase access to health insurance. The number of uninsured Americans has risen and the President is against any plan to make coverage universal, stating 'I'm absolutely opposed to a national health care plan.'

Most Americans are deeply unhappy with their health insurance – at least those with private health insurance. Medicare recipients, on the other hand, are actually happy with their system of socialized medicine. Or they were before the Medicare Modernization Act took effect. The passing of this legislation is an excellent lesson in modern civics and under-scores the willingness of many conservatives to lie to the public. The original idea was quite good – expand prescription drug coverage to seniors. That alone brought many people on board, but it got ugly when the details were allowed to trickle out. First, the act resulted in a 17.4% increase in premiums for people enrolled in Medicare, the largest increase ever. The administration was careful to release this information on a Friday afternoon before a holiday week-end to lessen the blow. If that were the only problem with this law, I might argue that it was an essential part of increasing coverage. But it wasn't.

The administration clearly lied to Congress and the public that the overall cost of the plan was estimated to be around $395 billion. But it was later revealed that the government's chief actuary for Medicare and Medicaid was threatened with being fired if he revealed that the administration had calculated the number to be about $156 billion higher. Why the lie? Well, many congressmen had said that if the cost went over $400 billion dollars they would not have voted for it. The price tag came in conveniently just below that. The actuary was told that he could talk freely with Republican members of Congress, but that he was not to talk to Demo-cratic members unless he cleared what he was going to say with administration officials first. Not to worry: it turns out he was wrong anyway. The total cost of the bill is not $550 billion after all. Instead, current estimates put it closer to $800 or $900 billion. (If lying to Congress isn't an impeachable offense, then I don't know what is.)

Don't worry; it gets worse. The system was set up so the Medicare program could not negotiate prices with drug companies to try to reduce costs through bulk purchase. No reasonable explanation for this policy has ever been offered.

In early 2006, the act went into full effect. Seniors now have to choose from scores of programs that are too confus-ing even for people involved in billing for health services to understand. Before the program went into effect, Medicare Prescription Drug Discount cards were made available to seniors by private companies. But these cards were fraught with fraud and were inexplicably complex to navigate, so only about 19% of eligible people actually signed up. Leading up to the implementation of the prescription drug benefit, the Centers for Medicare and Medicaid Services (the folks who oversee federal management of both programs) tried, but failed miserably, to explain to seniors how to navigate their benefits. This is not necessarily their fault: the plans are so complex and there are so many of them and the actual benefits to anyone enrolled are so obscured that it is almost impossible to figure out what you are getting for your money. Seniors are now widely reported to be choosing their policies at random and not being covered for the medication they need.

My sincere recommendation to those futilely trying to navigate their new options is that they send their papers to their Congressman or the White House and ask them to explain it for you. They passed the law; they are surely able to explain its consequences.

Medicare recipients aren't the only ones who have gotten the short end of the stick. The President and Republicans in Congress have shown absolute contempt for poor people who rely on Medicaid. While they have greatly expanded healthcare spending on seniors, have provided tax cuts and tried to get rid of the estate tax that even their own actuaries tell them are mainly going to aid the wealthy. They have done nothing to help out Medicaid recipients. The President actually proposed a $60 billion cut in Medicaid for fiscal year 2006. Luckily, Democrats in Congress were able to reduce that amount to $10 billion, but even that amount is completely arbitrary. No one has given any detailed justification for the cuts. It is a further indication that the moral hazard philosophy is guiding policy, and reflects contempt for socialized medicine.

The President does not understand healthcare

'I believe every American should have access to quality, affordable health care by giving consumers better information about health care plans, providing more choices such as medical savings accounts and changing tax laws to help more people, such as the uninsured and the self-employed, afford health insurance.'

Does any rational person believe that information and choices are what are keeping people from getting adequate healthcare?

'I believe we have a moral responsibility to honor America's seniors. Now seniors are getting immediate help ...'

The next day it was announced that Medicare premiums would be raised by the highest percentage in history; the prescription drug discount cards had 19% enrollment; and the new prescription drug benefit has caused widespread trouble in the retired population.

'The other issue regarding health care is whether or not health care is affordable and available. And one reason it's not in certain communities is because there is too many lawsuits.'

There are no credible economists that think that lawsuits have been a major reason for the dramatic increases in healthcare costs in the last six years. President Bush has repeated this sentiment over 70 times in speeches.

'We've got an issue in America ... too many good docs are getting out of business. Too many OB/GYNs aren't able to practice their love with women all across this country.'

Again this is in reference to lawsuits. Besides being funny, it is wrong. Doctors are not switching fields or going out of business because of too many lawsuits. The administration has yet to put forward a plan that both protects doctors from frivolous lawsuits and ensures a patient's right to be compensated.

'I mean, if somebody is practicing preventative medicine, it's going to mean Medicare costs go up. Medicaid costs will go up. Veterans' health benefits go up.'

This is absolutely untrue, except for the last bit. Preventative medicine will improve veterans' benefits. Every single professional medical society believes that preventative measures greatly reduce the incidence of expensive complications.

Some will say that there is a lot of corruption and fraud in the Medicaid system and that the cuts will force states to clamp down. There *is* a lot of fraud in the system, but there are ways to deal with fraud without reducing services to poor people. In fact, a reduction in spending will actually wind up increasing overall medical costs by decreasing the likelihood that people will get medical attention early on when sick or injured. When that leads to complications, hospitals will be forced to give free care to these people. This phenomenon has been shown to greatly increase the cost of services, which are passed on to other customers who have insurance. Either way, the system will pay, and it is widely accepted that catching problems earlier at the risk of more frequent medical visits is still cheaper than dealing with complications that lead to prolonged expensive care. Let's consider the facts.

Medicaid is the only health insurance held by 50 million low-income Americans. It pays for more than one-third of all births in the US and about one fourth of all American children are enrolled. It covers 7 million disabled people, and is the largest payer of HIV/AIDS care in the US. Finally, it pays for the care of 70% of all people in US nursing homes.

Whose luxurious healthcare services did the President conclude that the $60 billion in cuts, the equivalent of insuring 1.8 million children or close to 350,000 senior citizens, could be trimmed from? In the end, the dirty little secret that Republicans don't want you to know is that Medicaid and Medicare have been shown to be quite efficient in delivering what care they can, despite reports of corruption.

If the hub of medical cost increases is the laws and policies that regulate (or don't) the insurance agencies, then we have to consider the manipulation of that system that has contributed to increases in insurance premiums. I am rather simplistic in my view of systematic manipulation for profit without increased quality and access to care. It's fraud. I don't care if it's legal, it's wrong – and its time to put a choke collar on those involved or just remove them from the equation.

Rather than blindly cutting Medicaid funding, it would have been far wiser for Congress to pay for the development of computer code that they could supply to states to police the system for fraud. These programs have been used by private insurance companies for some time and are largely effective. Since the prices for services offered through Medicare and Medicaid are regulated, the main focus of fraud is the detection of billing for excessive services or services that are not performed. A *New York Times* investigation in 2005 revealed that the State of New York's Medicaid program might be losing as much as $18 billion a year to Medicaid fraud, much of it completely obvious fraud like a dentist billing for 900 procedures in a single day. That is approximately 40% of the state's total budget. A combination of computer algorithms and a handful of Medicaid police could save the entire system and allow expansion of services to more uninsured people without spending any more than is already allocated for Medicaid. This is especially true if those in charge of investigating are well trained, well paid and do not have to cut through tremendous bureaucracy to take action. Putting such a system in place on a national level would preserve services for needy people without draconian cuts that do not resolve any problems.

In the case of medical fraud, crushing criminal and civil penalties need to be put in place to make it so egregious that no one wants any part of it. This would include serious jail time and fines that do more than cover court expenses; they should be large enough to crush a person or company financially. Anyone committing large-scale fraud should be left just shy of destitute. Lawyers who actively participate in multiple malpractice cases that are proven to be flagrantly fraudulent should be disbarred. While patient advocates are against such solutions, there is no logical reason why those who deliberately try to capitalize through fraud should be allowed to prosper. There is a huge difference between these sorts of crime and crimes that are associated with poverty and cultural disadvantage, because they require advanced manipulation of the system and usually require many people to conspire. Rather than stocking our jails with poor drug offenders, why not fill them with unscrupulous white collar criminals and other parasites whose crimes have the potential to hurt far more people than some kid peddling marijuana?

After Hurricane Katrina devastated the gulf coast in 2005, rapper Kanye West caused a huge controversy when he said on national television during a benefit concert that 'George Bush doesn't care about black people.' Mr West was emotionally trying to make an important point: that the administration would have moved much faster during the relief efforts if white people were primarily affected. While I don't believe that the administration is full of abject racists, I do believe that Republicans in Congress and the administration, who have continually attacked Medicaid and Medicare while providing perks for the wealthy, have shown contempt for the poor. So, hoping that Mr West doesn't mind, I'll rephrase for him: George Bush doesn't care about poor people.

Mea culpa

Pointing out the importance of the doctor–patient relationship is a cliché, and a cursory look at medicine today will tell you that its value has been warped over time. At the heart of this relationship is trust. Patients are inherently vulnerable because of the uncertainty felt when asking another person for help.

The psychological game that gives patients the feeling that their autonomy has been stripped begins the very moment a patient seeks medical services. Choosing a doctor has been turned into nothing more than a roll of the dice. Most people have to designate a doctor as soon as they sign up for medical insurance. This often requires that a doctor be picked blindly from the insurance company's list of providers. Studies have shown that it is more typical for people to choose a new doctor than pay a premium to keep their old one. Insurance companies rarely allow you to have information on the educational background and expertise of a given doctor, and it is virtually unknown for them to supply you with malpractice history. So, while you have a choice, it is closer to choosing what's behind curtain number 2 than making an educated decision.

There is absolutely no incentive for insurance companies to provide thorough information on a doctor's background. In essence, their clients are treated as a commodity and it is their job to make sure that they make money from these commodities, not to provide a feeling of autonomy or choice, even though many of them will call their individual plans 'choice,' 'select' or 'premium.' But blame for not providing adequate information about providers cannot be placed solely on insurance companies. They have to retain doctors in their program, and one thing the medical establishment has fought strongly against is revealing information about doctors' backgrounds through insurance companies. So the general feeling is that it is the duty of the patient to ask doctors or to research it through state medical boards. Even then, it is extremely difficult to get detailed information about your doctor without jumping through multiple hoops.

Consider a typical visit to a doctor's office. You walk in and are handed a series of forms that require (or at least they ask) you to reveal all sorts of intimate information about your life. After you fill it out, you wait with strangers and read shitty magazines because other people have already stolen all the good ones. Half an hour after your appointment was supposed to start, you are escorted to a treatment room, raising your hopes of a rapid exit and recovery. The room is cold, sterile – hopefully in sanitary terms as much as in post-apocalyptic German decor – and you are told to disrobe, put on a paper gown that matches the design team's vision and sit on an uncomfortable paper-lined exam table under the unforgiving fluorescent light ... and wait. Here you might take the initiative to look through some drawers and cabinets for some interesting torture device to fiddle with or to bring home for the kids to play with.

The next person to come in will be a nurse or some other aide to take your pulse, blood pressure, breathing rate and other so-called 'vital signs.' The most accurate vital sign in my case is that I get up in the morning and fix myself a cup of coffee, but go ahead: take my temperature. By the way, I'm here because my eye hurts. After they leave, you wait *again*. Now you just want to steal something and curse yourself for not bringing a bag with secret compartments, or at least for not bringing the January 1991 *Time* magazine from the lobby.

Finally, relief arrives when the doctor comes in. He asks you to answer even more personal questions, many of which you have either already written down on the forms or told to the last person who was in the room. After about a minute or two, he scrawls something on your chart, maybe hands you a prescription for a medication and tells you the nurse will be back to take some blood. Panic! You explain that you have insurance and that the prescription will do just fine, but there is no changing his mind – you will get a needle as a token of his appreciation because the Jaguar parked outside wasn't enough. The time between the doctor leaving the office and the needle bitch walking in always seems to be the shortest waiting period in a doctor's office. After sucking it up and losing a bit of blood, you put on your clothes, walk out, give the secretary your ten dollar copayment and leave. The people in the waiting room will inevitably look to you for hope as you button up your jacket. As you leave, somehow you always say thank you, even though thanks to them is not at all what you feel. When you get back in the car, the first thing you always do is look at the band-aid on your arm and feel happy that it's over.

In any other business this sort of behavior would be con- sidered highly unprofessional. Why is it that when a mechanic works on your car you will question what they are doing, but when it comes to medicine we just assume that the doctor is right, even though we are questioning their activity in the back of our mind? What was that blood test for anyway?

The attitudes towards patients in doctors' offices are reflective of a larger medical culture, where establishing and maintaining the authority of treatment providers over patients is seen as essential. This attitude is a natural extension of the confidence that medical practitioners must have to do their job. A doctor without confidence in her ability is virtually useless. As for the office etiquette, it is thought by some in the medical community to be a necessary evil in a world where physicians' time and the high overhead of an office require doctors to spend less time with patients. Most doctors would much prefer to be able to spend more time with each patient and fully explain what it is that they are doing. But they have to overbook the office to ensure that they can see a constant stream of patients just to get by.

Some in the medical community would point out that their lofty salaries are compensation for crazy hours and that in the end they actually make less money per hour of training than almost any other profession. This is certainly true. But it doesn't make it right. On the surface, complaints from doctors would seem a lot of hooey from people who bring in a hefty profit. It really is hard to feel bad for people who have great respect in their jobs and an extremely high job satisfaction rate, and who make six-figure salaries. I'm with you on that one, but in fact doctors only represent about 2% of medical costs in the US. If you want to complain about salaries, take a close look at the seven-figure salaries of insurance company executives and hospital CEOs. Somehow, the healthcare industry, as well as just about every other industry, has allowed the disgusting escalation of salaries for lawyers and business people while the inflation rate of doctors' salaries has lagged well behind inflation for all other medical services and business salary rates.

One reason for the recent increases in medical expenditure is the fact that doctors and the medical community have been complacent in the area of treatment reform. However, solutions will have to come from within the insurance industry and the corporations that provide medical supplies and services before change can happen *en masse* in the treatment room. One thing is for sure: the overworked medical establishment and lack of time and attention paid to each patient have led to more mistakes, and there are things that can be done to limit mistakes, even within the current medical structure.

To err is medical

Any reasonable doctor will tell you that mistakes are common in medicine and misdiagnosis is not unusual. Every doctor makes mistakes. Some of those mistakes are minor, including the prescription of medication that will not help or which worsens a person's condition, and some are serious like amputating the wrong leg or inadvertently causing the death of a patient by missing an important detail or just plain screwing up. The public is largely unaware of how common medical errors are. Studies have shown that on average at least one mistake is made in dispensing medication every time you are admitted to a hospital. Granted, only 1% of the time does it lead to serious problems, but it is still alarming since many of those errors could be caught with the use of electronic prescription systems and institutional changes in drug dispensing.

Prescription mistakes are not just made in hospitals. They are one of the most common medical errors made by doctors and pharmacists. Most of those errors could be eliminated by electronic prescription systems, but doctors have not generally moved to them. Without pressure from malpractice insurance companies, or through changes in the law, it is unlikely they will. But prescriptions are just the tip of the iceberg. A large study from Harvard University found that approximately 4% of all hospital patients suffered serious ill effects from their treatment and that two thirds of those complications were a direct result of medical mistakes. Keep in mind that these serious mistakes ranged from an extended hospital stay to death. If we consider that the US spends over $1 trillion annually on healthcare, then even modest changes in protocols that reduce these mistakes could easily save billions of dollars, not to mention the pain, suffering and subsequent lawsuits.

In 2000, an Institute of Medicine study showed that the majority of mistakes made by doctors were the result of a lack of communication and of systems that can reduce or virtually eliminate some kinds of error. Subsequent reports have detailed how some doctors and hospitals have been slow to adopt minor changes in routine that can prevent medical errors. But the pressure is on, since a significant reduction in errors will result in an equally significant reduction in lawsuits and malpractice insurance. The first level of error reduction is the obvious stuff. For example, surgeons now routinely sign the arm or leg to be operated on during the presurgical consultation with the patient to prevent accidentally working on the wrong appendage. This seems like a stupid mistake to make, but when a surgeon is cruising through a day's worth of operations, a prep nurse dressing the wrong area or an X-ray being put on the viewer backwards can lead to a mistake. Making sure they are working on the right patient and that the person is not allergic to the antibiotic being prescribed are other examples where double and triple checking can make a big difference without soaking up needless time and energy with paperwork and bureaucratic regulations. It's simply making the 'double check' standard.

The next level of error reduction comes from recognizing that doctors are human and that humans make mistakes. Therefore they have to reduce the conditions that are known to increase the fallibility of humans. Many mistakes made by doctors are a result of unreasonable demands that they or their hospital make on them. The most obvious of these demands is a lack of sleep and breaks during the day. A 2004 study demonstrated that medical students who went no more than 16 hours at a time without a break or sleep made one third fewer serious errors than students who worked a traditional schedule that can require them to work for up to 30 hours in a row. The tradition of having medical students, interns and early residents work extended shifts has long been viewed as a rite of passage, but it has also been recognized that people working with little sleep will suffer in their performance at any job. This is why pilots, train conductors and even truck drivers now currently have limits set on the length of their shifts. Medicine has made some concessions in this area, but a sincere restriction on the work hours of all doctors, not just students, seems to be necessary for reducing errors.

The final level is the most costly and hardest to institute: eliminating the systematic problems that plague medicine. The biggest of these problems is communication. Most hospitals and doctors' offices are extremely busy, and communicating symptoms, treatments and procedure instructions within the system can easily fail. Here again, electronic records and a concerted effort to talk to patients and staff are vital to reducing errors. The more time a doctor spends with a patient listening to their symptoms and giving them stern instructions on how to continue their care after they leave, the better treatment will be. Systematic evaluation of procedure is key here as well. This is precisely what many medical services have done. These days you can go to facilities that specialize in one particular type of surgery or discipline. By doing so, doctors can form assembly line-like procedures to reduce error. Since they are more practiced at a few procedures rather than a wide variety, it is more likely that the doctor will know the ins and outs better than a generalist. This is not necessarily the best trend, since prices also tend to be higher at these specialized facilities, but the idea of regimented procedure and limiting practice seems sound. Also, the more familiar doctors are with specific drugs and the complications associated with them, the less likely they are to make mistakes in prescribing them.

The litigious craziness that dictates seemingly automatic payouts for any medical mishap, the fault of practitioners or not, has had a large impact on the cost of malpractice insurance, particularly in obstetrics. Make no mistake: this is passed right on to the consumer in the form of your own medical expenses.

In response to the exorbitant malpractice insurance costs, medical organizations have been lobbying congress strongly for legal reform that place a federal cap on putative damages in malpractice cases at $250,000. This is despite the fact that there have been several studies that indicate that these caps do not affect the number of lawsuits or the average damages awarded in malpractice cases. In short, these caps do not seem to be significantly lowering the cost of malpractice insurance. Nonetheless, many congressmen support such reform. What medicine on the whole has been slow to accept is that it is anger at the arrogance of many doctors that causes many people to sue their physicians. Don't get me wrong: our society is lawsuit crazy and there is a serious need for cultural change so that people start to accept that bad things happen and that legitimate mistakes can lead to personal suffering. What I mean is that fewer lawsuits will be filed if medical practitioners start to explain when, where and how things went wrong, and to apologize.

The argument against such changes in medicine has been that taking responsibility for medical errors in the absence of legal protection exposes doctors to more lawsuits. This is true, but there is strong evidence from institutions where this practice is mandated that talking to people about mistakes can significantly reduce the anger that leads to lawsuits. Several states have passed laws that prevent apologies from doctors from being admitted as evidence in malpractice lawsuits. This attitude has been extended in some states, which have passed or are now considering laws that require doctors to reveal medical errors to patients in a timely fashion, something that is not required and certainly not encouraged in most medical settings.

At some university-based teaching hospitals where apologizing has been instituted, the number of lawsuits and the overall liability cost have fallen by as much as 50%! The policies were put in place after significant research showed that patients were more likely to sue their providers if they felt that they had not been compassionate towards them. Some state laws seem to be going too far by not allowing admissions of guilt into malpractice cases. This is significantly different from an apology.

Only between 1 and 2% of medical mistakes result in lawsuits, and apologies could lead to opportunistic capitalization by those who think they are due. But medical practitioners have long recognized the value of reviewing mistakes. The problem is that this usually only happens in hospital settings and is done amongst other doctors, not with patients. Not only that, it is usually done in extreme cases, not with more minor mistakes. Talking about them and acknowledging them with patients will certainly improve care and possibly allow valuable lessons to be learned that can prevent future errors.

One of the biggest problems to overcome in reducing the number of malpractice lawsuits is to better prepare the courts to differentiate between an honest reasonable mistake and gross incompetence. Any intelligent person who has ever served on jury duty knows that they are in some pretty ridiculous company. Anyone who might actually have the ability to comprehend how medical decisions are made is quickly eliminated from the jury pool or manages to wriggle out of serving. This leaves a sympathetic pool of nimrods who see damages as a right rather than compensation and will sometimes award exorbitant damages because its on a large evil insurance company's dime. Studies have shown that patients' odds of prevailing in a malpractice suit depend more on their final physical outcome than whether they were actually subject to negligent medical care.

Serious legal reforms can be made to reduce flagrant abuse of medical malpractice liability and none of them should involve limiting the value of pain and suffering that an individual feels with arbitrary monetary caps. We should consider setting up courts with people who have medical expertise. I call it MedMal court (you can use that if you like). In essence, the court would act to triage medical malpractice by determining whether there is just cause for a jury to consider it further. Because the court would not involve a jury, but would consist of unbiased jurists with medical expertise, a lot of the frivolous cases could be shunted. There is precedent for such courts when dealing with cases within the vaccine industry. In 1988 the US started the National Vaccine Compensation Program to address the growing number of cases against vaccine manufacturers. While there are certainly a lot of problems with the vaccine court, the idea of using people with medical expertise to make judgments on injury is a welcome change. Unfortunately, that is where the similarity between these two ideas ends.

There is another issue that makes me quite queasy: doctors who are essentially professional 'expert witnesses.' There is a whole business around testifying in court whereby doctors are paid ridiculous fees to testify on behalf of a plaintiff or defendant. It is not atypical for a doctor to demand $10,000 a day and to be delivered to and from court in a limousine. This is just plain sad. I am not against doctors being compensated for testifying as a witness, but the industry has sparked the growth of charlatans who will say basically anything they are asked to in court. This has gone effectively unchecked with the exception of a few medical societies that review the testimony of doctors. But there are few or no repercussions against these leeches. Here again, an independent medical court where doctors can evaluate evidence would even the playing field.

In the end, the most pressing issue for a person to ask themselves is how to tell a good doctor from a bad one or a good hospital from a bad one. The answer, of course, lies in the definition of good and bad. It should be empirically obvious, but it isn't. If a doctor works in a risky field or just prefers to take difficult cases, his or her record might reveal a high number of patient deaths, complications or hospitalization. Attempts to quantify medical success on a doctor by doctor basis have largely failed because they tend to measure the wrong thing. There have been several attempts to collect information on death rates for hospitals. From 1986 to 1992 the government tracked and released the death rates of Medicare patients for every hospital in the country.

Media reports, and indeed the actual lists themselves, missed the point: that death rates at hospitals are a bad measure of success since they vary widely from year to year and are often a function of the types of case they take. For example, if a hospital has a large pediatric cancer ward, they might show a much larger than average rate of deaths for children than a hospital that has a small pediatric department and handles few or no pediatric cancer patients. What's more, these rates say nothing about mistakes or practices that are leading to deaths or whether they are leading to deaths at all. They also provide no information to hospitals about practices that can be instituted that would improve care. Noble as the sentiment behind these efforts was, they turned out to be a giant waste of time and only served to invigorate malpractice lawsuits and send people into a needless panic that they were not receiving adequate care.

Some people contend that improving care, reducing error and tracking the progress of doctors is too difficult to do and that the system is not organized to promote open access to performance. They are right, but many of the solutions to that problem are not difficult to implement. It all starts with record keeping. Most doctors still keep patient information in the form of paper charts and folders. This means that there is no reasonable way, short of sifting through mountains of paper, to get any kind of assessment of performance for the great majority of physicians. It also means that good practices that reduce error or increase efficiency cannot be gleaned from records. Take, for example, the fact that most people switch doctors every time they change jobs because of insurance enrollment differences. Very few people ever have their records forwarded to the next doctor. If medical records were electronic, an entire case history could be sent with a single email.

Not long ago, I changed jobs and thus switched insurance plans. I had gone through this process a few times, but that did not make it any easier. Like many people I was directed to designate a primary care physician within the plan. I had been working in Washington DC for a year and had not visited my previous physician in that time, but I figured I'd save some effort and use the same one. I went to the insurance company's website and sifted through the names, but could not find my doctor.

I always felt a little weird about picking a doctor at random, so I decided to put in a little more effort. Here, I ran into my first snag. The insurance web site did not provide any information about the doctors' backgrounds: not where they went to school, how long they had been in practice, when they were board certified or where they did their residency. It did not even tell me what medical degree my doctor held: MD or DO? The process was clearly not designed to allow patients to make informed decisions about who their doctor would be. In fact, it seemed designed to allow patients to pick their doctor solely on location.

Frustrated with the lack of information on their web site, I looked to the Virginia Board of Medicine web site. There, doctors who start a practice in the state are asked to give standard information including location of practice, telephone number, translating services, and percentage of time spent at location(s) of practice, but are also required to report information on their medical education, number of years in active clinical practice, all hospital affiliations, any academic appointments, professional publications, Medicaid participation *and* felony convictions as well as paid malpractice claims in the most recent ten years. This was a good start. I looked up the three doctors who were located closest to my home. Right off the bat I found that one had been convicted of dispensing medication outside of his practice. No details were given about the conviction, but it was serious enough that he had to surrender his medical license in the state of Massachusetts, where he was also licensed, and is barred from ever practicing there again.

When I looked in the paid malpractice claim section, I was faced with a clear explanation of the caveats of legal payouts, starting with the point that claims do not necessarily reflect wrongdoing. In this case, the doctor settled a suit in which a patient was diagnosed with colon cancer after a visit to his office. At a visit years earlier the patient had complained of rectal bleeding and nothing had been done. Since the information is self-reported, the doctor had the opportunity to explain that his notes from years earlier had not mentioned such a complaint. There is little or no oversight to the information given through the state board. No information about the actual case was given, just that there was a disagreement that resulted in the doctor settling the claim out of court.

I decided not to go with the doctor with the spotty record and chose a younger doctor who went to a good school and did his residency at a very reputable hospital. He had no convictions and had not settled a lawsuit according to the web site, but that doesn't mean that he is a good doctor. He is practicing outside of a hospital setting, so he is actually not held to any but the loosest of standards and is practicing unchecked. Had he made a series of prescription errors that, had they not been caught by an alert pharmacist, would have caused harm? Had he misdiagnosed diseases at a rate much higher than his colleagues? There is no way to know without going to them, and in that case you already need their help, decreasing even further the chance of getting answers. The fact is, the system is set up so that we spend more decision-making effort on ordering an entrée than we do picking a doctor, and that is just plain wrong.

A week later, I called my primary care physician's office to try to set up an appointment. As it turns out, he is part of a much larger practice and I could see any of the doctors there, but with a caveat. They had stopped taking my insurance six months earlier. I was told that they had been calling the company at least once a week for six months to be removed from their list of providers because they kept getting calls from people trying to set up appointments. The insurance company had not told their clients that their doctor no longer accepted their coverage and had actually continued to allow patients to choose doctors that no longer accept their insurance as their primary care physician.

The insurance company my employer uses is one of the largest in the country, serving over 8 million people, yet they value their doctors and patients so little that they cannot be bothered to provide information that allows them to make educated choices about their healthcare and allows their system to cause significant delays in treatment by not keeping their customers apprised of changes in their coverage. I chose not to name the company here because I don't want to point the finger at them specifically. This is a system-wide problem and has been reported for many companies. One would think that the government shouldn't have too step in and force these companies to be responsible, but it appears as though that is the only thing that will protect the public.

The Virginia Board of Medicine's site is considered one of the best out there, but they do not check the information on the site unless someone complains that a doctor's information is not accurate. It is kept current on the honor system and no doctors have lost their license because they didn't keep their records up to date. They do have a random audit process, but with tens of thousands of doctors in the registry, this is nothing more than window dressing on a lack of oversight. At the very least there should be a fixed system in every state detailing board sanctions and criminal activity. This system should be policed by the board, and failure to keep this information current should result in serious consequences. Again, repeat offenders should lose their license. Also, and I cannot believe that I have to say this, perhaps having doctors self-report criminal convictions, lawsuits and sanctions taken against them is not the most effective way to keep the public informed.

9

science education

"Scientists have made progress in their efforts
to design a more efficient politician..."

In case you've not worked it out yet, there is an underlying theme in this book: the public and the government do not understand science and medicine. That misunderstanding has led to widespread misconceptions about the way our bodies and environments work and a fear of progress. It is the failure of our education systems that has rendered the US scientifically stupid. Many scientists feel that this is a result of conservative policies and a recent phenomenon. On the contrary: this has been a trend in the US for decades as the public education system was left to rot. The conservative policies that are currently growing and causing anti-science rhetoric throughout society are a symptom of educational neglect, not the cause. It does, however, contribute to a feedback loop whereby growing misunderstanding leads to failures in science education, which leads to improperly trained teachers, poorly educated students, political and societal fear and further decay.

We're number 22!

The US consistently ranks well behind most other industrialized nations in science education, but it is clear that the situation is getting worse. All survey results on comparative education for high school science education place the US behind at least 16 other countries in the world. According to most surveys, we are even worse off than that. (The Third International Mathematics and Science Study placed the US at 16, but did not include many countries that are far ahead of us in science education, including Japan, Belgium and South Korea.) The most recent Program for International Student Assessment (PISA) study places the US at number 22! Either way, it's bad, very bad. Also, US high school students are well behind all countries that have significant research programs that compete against US scientists. We still hold sway over major scientific research findings in the world, but there is significant evidence that this is shifting. According to the National Science Board, 34% of all science jobs requiring a

List of national education rankings for science

1.	Finland	24.	Russian Federation
2.	Japan	25.	Latvia
3.	Hong Kong/China	26.	Spain
4.	Korea	27.	Italy
5.	Liechtenstein	28.	Norway
6.	Australia	29.	Luxembourg
7.	Macao	30.	Greece
8.	Netherlands	31.	Denmark
9.	Czech Republic	32.	Portugal
10.	New Zealand	33.	Uruguay
11.	Canada	34.	Serbia
12.	Switzerland	35.	Turkey
13.	France	36.	Thailand
14.	Belgium	37.	Mexico
15.	Sweden	38.	Indonesia
16.	Ireland	39.	Brazil
17.	Hungary	40.	Tunisia
18.	Germany		
19.	Poland		Source: PISA 2003
20.	Slovak Republic		
21.	Iceland		
22.	**United States**		
23.	Austria		

PhD in 2000 were filled by people who were born outside the US. That number wouldn't seem so startling if it weren't for the fact that in 1990 that proportion was only 24%. Similarly, in 2000, 58% of all post-docs in the biological sciences in the US were on temporary visas and in 2001, 35% of all PhDs awarded in science and engineering went to students from foreign countries. If you just look at math and computer science, foreign students made up 49% of all recipients. Engineering was worse, with 56% of PhDs going to foreign students. By the way, your tax dollars are helping to fund a lot of these programs, as graduate students in the sciences do not usually pay tuition and are paid a stipend to do research.

The combined evidence of high school performance, awarding of advanced degrees and job placement indicates that the US is quickly losing its edge and is producing high school graduates who are scientifically illiterate. Next time you crack a joke about the Brits or the French, remember: their children are smarter than your children. Not fun to think about? Well, while you are waving your foam 'We're number 22!' finger in the air and chanting 'USA! USA!,' think about this: according to the National Science Teachers Association, only 26% of high school graduates from 2003 can be expected to have the knowledge base to be able to complete a college-level science course. We're not talking getting As or Bs; we're talking passing the course; and trust me, these courses are not taught in a manner that requires them to think critically and truly understand science.

The failure of US high school science teaching has had a caustic effect on the higher education system. The great majority of first-year college science courses do not require understanding of any complex concepts. They are largely exercises in memorization that cover much of the material that is in high school science courses. When one considers the fact that a college education costs approximately $11,000 per year for a student to attend a public university for room, board, tuition and fees, and that it costs $27,000 per year to go to a private university, the fact that students are learning material that they were supposed to have mastered at the high school level should be infuriating. But you don't hear them shouting about the way high school science is taught; you don't hear them at school board meetings screaming about the fact that their science teachers do not have degrees in science or that the curriculum is outdated and often arcane.

What gets the hair on the back of parents' necks to stand up? Grades. There is a growing trend in the US for parents to fight for higher grades for their children without regard to their actual class performance. It is estimated that approximately 25% of all US high schools are involved in at least one lawsuit each year. That is almost double what it was ten years ago. While many of those suits have to do with personal injury or other legitimate cases of negligence, there is an exponential trend of parents threatening legal action over their child's performance in the classroom. The majority of these cases never see the inside of a courtroom because spineless administrators cave in to the pressures of these psycho-parents and will actually force teachers to give students a higher grade than they deserve. Often, administrators actually change the grades of students without notifying teachers, or worse, after teachers refuse to. This is particularly true for tenured teachers, who can actually fight the issue without fear of losing their jobs. Non-tenured teachers have been fired after protesting such changes. So, more often than not, they just give in to school administrators or parents. The reality is that any administrator or teacher who changes a student's grade because of a complaint without sufficient academic performance to justify it should be fired!

The authority of teachers has been eroded away as parents use what used to be a last resort course of action reserved for only the most extreme of cases. The arguments are almost never based on their child's actual performance in a class, but on grounds that the teacher did not treat their child fairly. Parents increasingly care less about how much or what their child learns and now focus on grades; dumb, rambunctious or lazy child be damned. Sparing the rod is now a national pastime.

Worse, teachers actually give lessons on how to beat standardized tests without knowing the material. Sometimes this is because the questions are poorly worded or the tests are inherently flawed, but in other instances it is just to boost grades artificially without requiring the students to know any more of the curriculum. In addition, the tests given are almost always multiple choice questions. This should mean that the questions have to be straightforward so that a student can understand the concept that is being asked. But without fail, you will find questions on every science test that are designed to trick the students into giving a wrong answer. To get these questions right, the student doesn't just have to know the material, they have to be skilled test takers. In other words, students that know the material will get the question wrong. This is just plain bonehead logic. The focus on test taking rather than finding a true test of a child's knowledge base is a growing misguided trend in schools, fueled by overcrowded classrooms, under-funding of schools, and standardized multiple choice-style testing that forces teachers to be automatons who can rarely put the creativity or time into teaching students new science or getting them to understand how science happens. The subject matter might as well be a string of word associations, because students are scarcely required to actually understand the most basic concepts.

The best example I can point to is that a student can pass high school biology in every one of the US states and go to college without knowing that genes are made of DNA. This is insane.

One of the best recognized ways for students to learn science is by doing science in lab classes. If you think back to your high school biology class, you are most likely to remember the lab work that you did. Whether it be dissecting a frog or spooling DNA onto a glass rod, labs utilize a hands-on approach to science that is difficult to forget. Now students are increasingly likely to simply watch a teacher do the experiments in front of them. The rationale is that it is logistically difficult to get students to complete labs and the equipment is expensive. If labs are taught and students do participate, classes are designed for quick and easy experiments that don't challenge students to pay attention and work diligently. Teachers call these 'cookie cutter' or 'cut and paste' labs reflecting the lack of challenge involved in completing them.

Many are an extension of memorization exercises where students simply look through a microscope at a slide and draw what they see. This is hardly science and hardly useful for learning.

Take the New York State curriculum. New York has had standardized final exams for decades in many subjects, called Regents Exams. The state has the entire spectrum of education problems, from urban classroom overcrowding in poor neighborhoods to rural classrooms with only a handful of students. Yet the state ranks above the national average on the National Assessment of Educational Progress, also known as the 'Nation's Report Card,' conducted by the Department of Education. That doesn't necessarily mean that students who graduate from New York State high schools with a Regents diploma are well educated in science. The state has tried to revamp its science programs, but wound up with a lot of the same tired old material that doesn't really convey the basic principles necessary for building a sound foundation for college – or life for that matter. In their efforts, they even renamed the standard biology class 'The Living Environment,' which stresses ecology and is not much different from the science taught in the 1950s. That's right, students in New York no longer take biology; they take Living Environment. Puhleeease!

Here's Question 3 from the June 2005 exam for high school students (see opposite).

After staring at the diagrams for a while you probably figured out that one diagram was designed to represent a cell and the other a very masculine woman. You probably started looking at the possible answers and were somewhat perplexed because the answer didn't seem obvious. Well, the actual answer is 4, the kidney and the cell membrane. Don't worry if it still doesn't make sense to you. I have a PhD in genetics and I still don't get it. The question is supposed to test a student's ability to assess the diagrams, know what the different structures are and know what they do. I guess the

Which structures in diagram I and diagram II carry out a similar life function?

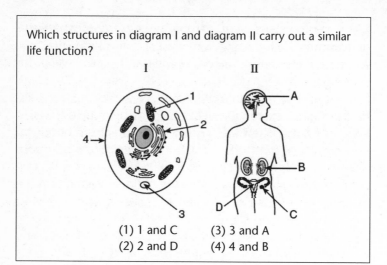

(1) 1 and C   (3) 3 and A
(2) 2 and D   (4) 4 and B

link between the cell membrane and the kidney is supposed to be that they both filter material, but this is neither a useful nor relevant comparison for understanding what either really does or for understanding comparative anatomy, if that is in fact what they were trying to test. The point is that the comparison is kind of stupid and a student who knows the diagrams and all of the components well might get it wrong.

Let's take another example, question 42.

Oak trees in the northeastern United States have survived for hundreds of years, in spite of attacks by native insects. Recently, the gypsy moth, which has a caterpillar stage that eats leaves, was imported from Europe. The gypsy moth now has become quite common in New England ecosystems. As a result, many oak trees are being damaged more seriously than ever before.

Certain insects are kept under control by sterilizing the males with x rays so that sperm production stops. Explain how this technique reduces the survival of this insect species.

This question seems deliberately long and complex to get to a simple point, Sterile animals can't reproduce. Duh! Where is the science? After a year-long class are there any students that don't get that concept? I would hazard that most of the students that get this question wrong are actually confused by the wording of the question itself, not the concept that sterile animals can't make babies. Question 31 is even weirder.

An experimental setup is shown below.

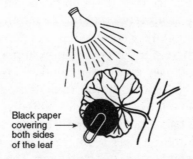

Black paper covering both sides of the leaf →

Which hypothesis would most likely be tested using this setup?
(1) Light is needed for the process of reproduction.
(2) Glucose is not synthesized by plants in the dark.
(3) Protein synthesis takes place in leaves.
(4) Plants need fertilizers for proper growth.

Besides having an experimental setup from the Stone Age, it is absolutely ridiculous to ask a question like this because not every student that takes the test is required to do this experiment and most are not even shown it, so it is a pretty pointless question. No one will ever need to have skills that require them to look at a foreign experimental setup and divine what they are trying to test ... unless, of course, aliens set up camp here and start designing experiments on plants. To get this question right, a student can just be good at taking tests and use a little logic. Without any results from this antiquated experiment, the student can hazard a guess that light has something to do with it: the giant light bulb and dark paper kind of give that away. With just two answers to guess from that involve light, the average person could guess that the answer is 2 without knowing anything about the experiment or glucose production in plants. All you have to know is that leaves don't have anything to do with plants making more plants. A little test-taking skill should never be able to substitute for knowledge.

It gets worse. The test is graded on a curve, so that in 2005 and 2006 a student only had to get 39 out of a total of 85 points on the test to pass. That's 46%! Believe it or not, New York State actually lowered that number from 2004, when a student had to get 48% correct to pass. I don't want to pick on New York here. I use them as an example because their curriculum is actually considered good and New York State students actually tend to go on to higher education more frequently than a lot of other states. Often you hear states' administrators talk about 'raising the bar' for science education. These administrators seem to be out of touch: the schools are playing limbo, not high jump. The science curriculum in every single state is antiquated and does not adequately cover modern-day science. They need to change the game, not the standards for the one they are losing now. Calling the situation dire at this point does not seem too extreme. So what are we to do about it?

All children left behind
Unfortunately, attempts to correct the situation at the federal level have resulted in lower curriculum standards and have provided little guidance for schools to use to actually improve curriculum. Current federal policy has fueled neglect of science education and has added lemon juice to the overall education cuts in the US. In January 2002, President Bush signed 'No Child Left Behind' into law. The law is fairly complicated, but in essence the goal is to raise the academic standards of all students through standardized

testing, school accountability, and raising the qualifications of teachers. The program was rushed into practice, starting with standardized testing of math and reading. The law allows for students attending schools that are consistently underperforming to transfer into another school in the district unless that school is full. If there is no space in any schools that are performing well, then the school district is to pay for after-school programs and tutoring for any child who wishes to receive it.

'Wow, that sounds great. What an inspiring showing of leadership.' Not so fast. Like most other policies that the Bush administration has put forward, it is all about window dressing for an easily influenced public. Science tests are not to be introduced into schools until the 2007–2008 school year, but unlike the math and reading tests, science tests will not contribute to the main measure of a school's progress, called the Adequate Yearly Progress score or AYP. This means that the legacy of undervaluing science education is now law. If a high school fails to reach an adequate AYP two years in a row the students can switch schools. But science performance is not factored into the calculation of AYP, so a school can have the worst science program in the country and still pass or even score highly with its AYP.

Plus, in his 2006 budget request, President Bush called for funding that is $9 billion dollars less than the act authorizes, so schools won't be able to pay for the extra tutoring and after school help that is mandated by the law. In 2003, the National Governors Association voted to label No Child Left Behind an 'unfunded mandate.' This means that they recognize it as a demand that the federal government has put on the states but not supplied enough money for. The Utah state legislature passed a law that puts the state's education priorities ahead of the No Child Left Behind priorities in the first public embarrassment of the administration's bungled law. In addition, the National Education Association, along with Michigan, Vermont, Connecticut and Texas are suing the government to be relieved of any requirements of the law that the government is not paying for. In response, former Bush administration education secretary Rod Paige called the NEA a 'terrorist organization.' Oops.

No Child Left Behind also misses a major problem with the funding of public schools. For most states the mainstay of public school funding is local property taxes. So, if you live in a wealthy area, you likely have a house that is valued higher than one in a poor neighborhood. That means that you pay higher property taxes and your school district gets more money as a result. Thus school districts in wealthy areas get more money than those in poor areas and their school gets to hire well-qualified science teachers and can afford a wide array of science equipment for the classroom. It is possible and not uncommon for two school districts to be right next to each other and have radically different funding because of their tax situation. Not coincidentally, this also sets up a racial barrier in some areas, where African American and Hispanic communities tend to be less well off than neighboring white communities. If you live in a poor area, well, you are shit out of luck. And if a school does not receive an adequate AYP score two years in a row, then it loses its federal funding. So schools in poor neighborhoods, which are traditionally the ones in trouble, will remain that way.

The very essence of No Child Left Behind is that teachers are to blame, not under-funding of school systems. Granted, there are a lot of bad or unqualified teachers in the nation's school systems, but merely demanding that schools get better without providing the funding for them to do so will be a colossal failure. The AYP itself has drawn tremendous criticism from teachers, school administrators and state officials. It is based on 37 measures of a school's performance. If a school hits 36 of the measures, it is just as much a failure as one that hits none. Also, schools are not evaluated for their improvement. So, if a school goes from meeting none of the criteria in one year to 30 the next to 36 the following, it still loses its funding, even though it has shown remarkable improvement in just two years.

This all bodes very ill for science education. Since science testing is not part of the AYP, it will be left behind. Students will continue to perform poorly and parents will continue to pay huge sums for their kids to take what amounts to remedial science classes at the college level. There is no incentive for schools to boost their science departments; in fact, with the strict testing rules placed on schools, the focus has shifted away from it and thus our education system will continue to produce adults who have little or no knowledge of common scientific facts or the very basics of biology. This leaves people poorly equipped to deal with health problems and discourages them from going into science and medicine. In the mean time, the Bush administration has been busted for secretly paying journalists big bucks to praise No Child Left Behind, a practice that the Government Accountability Office has called illegal. In a particularly flagrant move, conservative commentator and syndicated columnist Armstrong Williams was paid $241,000 to praise the law and the administration's work, but never disclosed that he was a paid spokesman for the White House in his columns or on-air appearances. Somehow the public doesn't have a problem with this sort of propaganda.

All of the pressure put on states to improve school performance has led to creative, but egregious, reporting of high school graduation rates. So not only is the US lagging behind the rest of the world in science education, we are lying to ourselves about how well our students are doing overall. One would think that calculating the graduation rate would be a pretty easy. It should be an intuitive function of the number of students that graduate compared with the number of students that attend high school. But rather than be honest about how the education system is failing, states have opted for creative accounting.

In 2005, The Education Trust, a non-partisan education advocacy group, released a report detailing the situation. It showed that New Mexico's 90% graduation rate is a result of only counting the percentage of seniors who graduate, leaving out the students who drop out between freshman and junior years. Even worse was North Carolina, whose 97% graduation rate was actually the percentage that received a degree in four years or less. This means that all students who drop out are excluded and that 3% take more than four years to graduate. So if 99% of students dropped out, but 1% graduated in four years or less, North Carolina would have a 100% graduation rate.

Under the No Child Left Behind policy, states must show progress in bringing students up to state standards for reading and math and at the same time must make progress towards improving graduation rates. That pressure has clearly contributed to the situation, and many states have set ridiculously low targets to show improvement. South Carolina and New Mexico decided that as long as their graduation rates don't go down, they have made enough progress to satisfy the provisions of the law. The department of education gets a big fat 'F' for failing to provide any leadership in science education or in demanding that states use an honest formula for calculating performance and graduation rates. It's all a big lie, and it means that science education is not going to improve unless the public stands up and demands it.

But don't take my word for it; listen to what the students have to say. In 2005, the National Governors Association released the results of a survey of 10,378 high school students on how they felt about their education. Astonishingly, of the 89% of students who said they intend to graduate, fewer than two thirds felt that their schools had done an adequate job of preparing them to think critically and analyze problems. Similarly, approximately two thirds said they would work harder if their classes were more challenging or more interesting, and fewer than two thirds felt that their schools had done a good job of preparing them for college or challenging them academically. Even those who dropped out cited the fact that they were not learning anything as the top reason for their decision. Only one in nine cited the work as being too hard as the reason for their decision not to continue.

Unfortunately, the evidence of failure is not just seen in the dispassionate attitudes of students, it is all around us.

Evolving controversy

The best modern example of unfounded fear and hatred of science, the most clear-cut example of ignorance of science in action, is the hubbub over teaching evolution as 'just a theory' in schools, alongside 'alternatives.'

Before we get into details, let's be absolutely clear on this: there is no controversy over the scientific principles of evolution by natural selection within the scientific community. It is the foundation of modern biological sciences, including the study of genomes, molecular biology, neuroscience and every other branch of biological research. The evidence is mountainous and absolutely irrefutable. Anyone who says otherwise is either lying or wrong. Simply put, the principles of evolution are essential to understanding biology. Without them, modern biological research, not to mention much medicine and conservation, would not be possible. Yet almost 150 years after Darwin first posited this explanation for life's rich tapestry and 80 years since the infamous Scopes Monkey trial first brought the issue to the courts, controversy is flaring up across the US once more.

There are many kinds of creationism and many of them have evolved (ahem) as a reaction to scientific evidence for evolution. While for our purposes it is not important to go through them all, it is important to recognize that, unlike evolution, there is no decisive creationist doctrine and no cohesive movement that represents all brands of creationism. The different types range from Flat Earth creationists, who believe the world is, well, flat because the Bible mentions the four cor-

ners of the Earth, to more progressive forms of creationism
that recognize that the Earth is billions of years old and that
there has been some evolutionary change, but that God set
the ball rolling with a set design in mind.

What they all have in common is that they invoke God to
explain the natural world and that they believe there has been
a vast anti-Christian conspiracy amongst non-Christians and
scientists to undermine their faith. Either way, the issue boils
down to teaching faith in science classes rather than teaching
proven science. The same people who want creationism
taught as an alternative to evolution would be up in arms if a
Muslim community announced that they would teach from
the Koran in science classes.

If you ask a creationist if there is anything that could convince them that strict creationism does not provide a rational
explanation of the real world, the answer is invariably 'No.'
Thus, rational discourse over scientific findings that make evolution as essential as teaching that genes are made of DNA
falls apart through rigid unscholarly rhetoric based on religious extremism. This has prevented most fundamentalist
Christian children from learning that there is nothing in the
theory of evolution that really threatens the Christian faith.

During the 2000 Presidential election, then Governor Bush
stated that:

> I'd make it a goal to make sure that local folks got to
> make the decision as to whether or not they said
> Creationism has been a part of our history, and whether
> or not people ought to be exposed to different theories
> as to how the world was formed,

in response to questions about the Kansas State Board of Edu-
cation's decision to eliminate evolution from the curriculum.
Then the scientific and education communities cringed as the
Gore campaign, through a spokesperson, said that the Vice
President favored teaching evolution, but that 'Obviously,

that decision should and will be made at the local level, and localities should be free to teach creationism as well.' This represented a cave-in to religious interests rather than standing up to tyrannical extremism. Soon after, the spokesman clarified his position, having checked into the 1987 Supreme Court decision in Edwards v. Aguilar. This found that the Louisiana statute that prohibited the teaching of evolution unless creationism was also taught was an unconstitutional endorsement of religion. He clarified his stance by stating that creationism should be taught in the context of theology or religion classes, but the damage was done.

The Christian Right has danced like a prize fighter on this topic by adjusting their arguments to suit court rulings and new research. After they failed to remove evolution from the high school curriculum and failed to have creationism taught in science classes, they came up with a new tactic. Rather than try to contradict the science, they decided that if evolution had to be taught in classes then the limitations of evolution must be taught as well, and that creationism filled those gaps well. Better still, they came up with their own 'scientific' field that could be 'researched' as an alternative to evolution. They call it 'intelligent design,' or ID.

Briefly put, intelligent design states that the world is too complex to be explained by natural phenomena and that there must therefore be a supernatural force guiding the creation of humans, the Earth and all life forms. Their main target is to attack the teaching of evolution, which simply states that the great diversity of life forms occurred as a result of gradual mutation and adaptation. The best part is that they rarely mention the words 'creationism' or 'God', just some mysterious force acting to make everything just so (wink wink). This departure is an outright recognition that teaching theology will have to be done by stealth. It is common not even to use the word 'evolution,' but instead to refer to Darwinian theory as part of a greater conspiracy of science to explain the natural world that they call 'materialism.' It is a carefully orchestrated public relations campaign that is doing serious damage, not to science, but to the public education system, subjugating serious students and teachers to the will of the church. Intelligent design is creationism with a web site.

The driving force behind the movement is the fundamentalist Christian Right empowered by 'research' findings produced through the Seattle-based Discovery Institute. Coincidentally, this was formed just three years after the Supreme Court shot down the Louisiana statute on teaching evolution and creationism together. It is a private non-profit entity that gets the majority of its funding from 'foundations,' private donations and membership fees.

Marketing God

In a startling revelation of the truly unscientific pursuits of the institute, an internal document from the institute made its way into the public domain in 1999. Called the Wedge Document, it outlines a grand scheme to replace evolution 'with a science consonant with Christian and theistic convictions.' They use the term 'wedge' to describe their philosophy and tactics, stating in the document that

If we view the predominant materialistic science as a giant tree, our strategy is intended to function as a 'wedge' that, while relatively small, can split the trunk when applied at its weakest points.

Indeed, the document compares teaching evolution to Marxism, blames it for the unraveling of society and accuses biologists of promoting social relaxation of personal responsibility for one's actions.

This is a similar argument to the one made by former Senator Rick Santorum from Pennsylvania, who blamed academic liberalism in the Boston area for creating an environment with relaxed moral standards that led to children being sexually abused by Catholic priests. (Gee, I can't understand why he lost his Senate seat in the 2006 election.) This sort of warped view can also be seen in their plans to spread intelligent design into the UK, Israel and other scientifically influential countries, but its success in the US is not likely to be matched abroad. Those involved can be accused of immoral misrepresentation of science, of being dogmatic and perhaps divisive, but not of being unskilled in their tactics to undermine the education system. It's important to understand the depth of the plan in their own words.

The Wedge Document outlines a three-phase plan including 5-year and 20-year objectives for the institute. Phase I includes the creation of scientific research, writing and publicity to try to promote intelligent design as a legitimate scientific concept within the scientific community. I'll tear into that later. Phase II aims to increase publicity in the general public. To do this they aim to 'cultivate and convince influential individuals in print and broadcast media, as well as think tank leaders, scientists and academics, congressional staff, talk show hosts, college and seminary presidents and faculty, future talent and potential academic allies.' Phase II also aims to 'build up a popular base of support among our natural constituency, namely, Christians.' They intend to do this 'primarily through apologetics seminars.' These are essentially training sessions where creationists are taught how to defend intelligent design from those more educated than they are about the scientific evidence for evolution. Phase III is aimed at cultural confrontation and 'renewal.' They plan on doing this once they have built an army of believers who can help them 'move toward direct confrontation with the advocates of materialist science....' This will not be done through any sort of legitimate scientific discourse. They plan on going so far as to 'pursue possible legal assistance in response to resistance to the integration of design theory into public school science curricula.' They state very plainly that they will attack the 'specific social consequences of materialism and the Darwinist theory that supports it in the sciences.'

This is scary stuff that sounds like a cult, and while the chances of them successfully turning intelligent design into the 'dominant perspective in science' are slim to none (because of the flimsy evidence they present and the unimpeachable strength of evolutionary scientific data), they have been very effective in being taken seriously in the education system. There are now close to 30 states that have had alternatives to evolution proposed at the state or local education level. It is this success that has endowed the institute with a healthy dose of hubris.

Their response to the allegations of divisive tactics that have little resemblance to legitimate scientific endeavors and which sound more like a campaign to promote a boy band was to put out a report entitled 'The "Wedge Document": So What?' where they spend 19 pages talking about how legitimate their efforts are and how they aren't pushing theocracy. A word to those who have been indicted by their own underhanded policies; if you need 19 pages to backpedal from your own six pages of policy, then 'So What?' doesn't really capture your own attitude towards you being publicly exposed. The 'So What?' document didn't really repair the damage done, but neither document seems to have greatly affected the stance of the religious right, who welcomed intelligent design as an equal to real science without really understanding either.

At the heart of the matter is the simple fact – watch my lips: FACT – that there is nothing formally testable about creationism or intelligent design. It is similar to the idea that lightning happens when Zeus throws down thunderbolts or that the Sun moves in the sky when Apollo drags it on the back of his chariot. You can believe it if you want, but you can't test it. Evolution on the other hand? Just ask a racehorse breeder, or a rice farmer or a malaria nurse – they get to test it, to watch it in action every day.

Excerpts from the Wedge Document

The proposition that human beings are created in the image of God is one of the bedrock principles on which Western civilization was built. Its influence can be detected in most, if not all, of the West's greatest achievements, including representative democracy, human rights, free enterprise, and progress in the arts and sciences.

Yet a little over a century ago, this cardinal idea came under wholesale attack by intellectuals drawing on the discoveries of modern science. Debunking the traditional conceptions of both God and man, thinkers such as Charles Darwin, Karl Marx, and Sigmund Freud portrayed humans not as moral and spiritual beings, but as animals or machines who inhabited a universe ruled by purely impersonal forces and whose behavior and very thoughts were dictated by the unbending forces of biology, chemistry, and environment. This materialistic conception of reality eventually infected virtually every area of our culture, from politics and economics to literature and art.

The cultural consequences of this triumph of materialism were devastating. Materialists denied the existence of objective moral standards, claiming that environment dictates our behavior and beliefs. Such moral relativism was uncritically adopted by much of the social sciences, and it still undergirds much of modern economics, political science, psychology and sociology.

Materialists also undermined personal responsibility by asserting that human thoughts and behaviors are dictated by our biology and environment. The results can be seen in modern approaches to criminal justice, product liability, and welfare. In the materialist scheme of things, everyone is a victim and no one can be held accountable for his or her actions.

Finally, materialism spawned a virulent strain of utopianism. Thinking they could engineer the perfect society through the application of scientific knowledge, materialist reformers advocated coercive government programs that falsely promised to create heaven on earth.

Discovery Institute's Center for the Renewal of Science and Culture seeks nothing less than the overthrow of materialism and its cultural legacies. Bringing together leading scholars from the natural sciences and those from the humanities and social sciences, the Center explores how new developments in biology, physics and cognitive science raise serious doubts about scientific materialism and have re-opened the case for a broadly theistic understanding of nature. The Center awards fellowships for original research, holds conferences, and briefs policymakers about the opportunities for life after materialism.

The cornerstone of ID's smoke and mirrors act is a clever piece of semantics. In describing evolution as 'just a theory' they play on the difference between our everyday usage of the phrase 'a theory' to mean 'a guess' and scientists' use of 'a theory' to mean explanations of generalizations (laws) based on observation and experimental tests which confirm a hypothesis. Adding to this confusion is the common public misconception that science works as a clear hierarchy going from hypothesis to theory to law. *Not true!* For example, we have laws of gravity, but no one main underlying theory on gravity. In other words, we cannot explain why exactly one body draws itself to another. There is no law of evolution because evolution *is* the actual explanation for changes we see over time. Creationism, in a sense, can be called a hypothesis based on religious text.

This is why when a Kansas school board decided to consider intelligent design, no scientists showed up to the meeting, (a) because it was not worth their spit to discuss it; and (b) because it was pretty clear that the board were going to decide to teach religion in science classes no matter what scientists said. It most certainly wasn't because they felt 'threatened by the "science" of intelligent design.' Researchers take the tack with ID they do with all new ideas – what are the claims? What is the evidence? What does the data look like? With intelligent design, the data does not even rise to the level of having a debate – as the physicist Wolfgang Pauli would have said, it's 'not even wrong.' Any other idea in science with data this weak would also be resoundingly ignored or admonished, especially if there was nothing in it that was formally testable and if it flew in the face of overwhelming data.

Scientists in all fields will continue to rely on evolution as a central tenet. Those who specifically study evolutionary biology consider the strength of intelligent design to be little more than a running joke. Since intelligent design proponents generally operate among the public rather than choosing to engage the scientific community, they have been left unchecked by the minds who could crush their ideas if presented in any serious academic arena. They know this, and will therefore continue to operate outside the scientific community. To date there has been only one peer-reviewed published paper on intelligent design, and it appeared in a journal that exceedingly few working scientists read or even know exists. Nonetheless, this has given the movement firepower to say that it is a legitimate theory because it was published in a peer-reviewed journal, even though the paper did not contain any new experimental data that supports the theory.

Poorly designed

Intelligent design is not (ostensibly) based on a literal interpretation of the Bible, as many people believe. Rather, it is based on assumptions about the limitations of evolution and mathematical modeling of randomness. Let's start with the biological arguments for intelligent design, which are broken into two basic parts. The first claim is that the unexpectedly intricate and complex makeup of every cell could not have been created by a random process because so many of the parts are essential. The second argument is that so much in biology is what they call 'irreducibly complex.' This means that if you remove any part of a structure or functional group of proteins, the process it is involved in won't work, so there is no way it could have evolved through a process of minimal changes to the individual parts.

While it is absolutely true that legitimate scientists cannot and have never claimed to be able to reconstruct how early cells worked, they are now working on breaking down the minimal units required in a cell. However, most scientists don't really consider the issue of how life started or even how the first cells were formed. This is simply because there are plenty more tractable problems in biology to tackle, such as how a memory can be formed, what causes disease or how the immune system works. We don't have a good idea of what the Earth was like 3.85 billion years ago when life is thought to have started. Think of it this way. We have trouble figuring out what life was like for humans several thousand years ago, so imagine trying to figure out what it was like for the first living cells billions of years ago. True, scientists cannot tell you what the first cells might have looked like or how the complex parts of a cell evolved or how the first animals came about or what they had inside them. This is simply because the physical evidence of early life forms has, not surprisingly, decayed over several billion years.

The theories of how a cell came to be are taught, but these are based on experimental evidence showing how certain molecules like amino acids can form under the right conditions. The fact that biologists cannot give precise details of how all life came to be is the very foundation of the intelligent design principle. That life no longer exists in these forms hardly means that there is evidence of a creator and certainly does not mean that simpler forms of life did not exist. It was only a few years ago that most scientists believed that humans would have close to 100,000 genes, but we now know that the number is closer to 25,000. We are still in a great age of discovery – and that is the exciting thing to be teaching our youth.

If classrooms were to start teaching all alternative explanations for the origin of life, the entire class would be consumed by speculative ideas and would blur the line between facts about biology which are scientifically proven and religious or metaphysical ideas. To question evidence is certainly to be encouraged, but learning science absolutely requires that such lines of question come from rationality, not scripture. That is best left to classes on theology, where such perspectives can be taught in the context of doctrine and juxtaposed against alternate views. Science classes must be based on empirical scientific evidence. Religious belief is fundamentally different from scientific debate.

The idea that certain biological units are 'irreducibly complex' has been touted by intelligent design proponents as absolute evidence that their theory is valid. It came to prominence in a book by a Senior Fellow at the Discovery Institute, Michael Behe, in his book entitled *Darwin's Black Box*. The examples that are commonly rolled out are the seemingly impossible evolution of an animal's eye, the flagella on sperm and many micro-organisms, and blood clotting. Simply put, if any one component of any of these systems is removed, then it doesn't work any more. ID supporters therefore claim that they could not have evolved. Bring this argument to the scientifically bankrupt Christian Right and they will cheer for the man who has proved that God created the Earth in seven days. But this theory has taken a stern ass whooping by others of a more broad-minded bent. Irreducibly complex systems can arise in several ways.

I like biologist Allen Orr's explanations best. First, there is the concept of cooption, where molecular parts evolve for one reason and then are incorporated into a new function. This has been shown for many proteins that contribute to irreducibly complex systems. Today, global positioning systems are common in cars, and it would not be surprising if fifty years from now cars are driven by this type of system without manual steering systems. Looking at that car of the future, you would not be able to determine that the technology was adapted separately and slowly integrated into the car. The irreducibly complex system also does not consider the fact that if some part were added to a system that improved performance and added an advantage, it would not necessarily be essential. As more parts and modifications to the system are added, that original addition becomes essential. Blood clotting, which requires close to 20 proteins, is the result of another mechanism in evolution, gene duplication. In that case, you get two copies of the gene, which gives them more freedom to evolve because one can compensate for the other. The proteins involved in blood clotting are quite related to each other, indicating that there were several rounds of duplication and subtle evolution to improve clot formation. There are many other mechanisms that can contribute to the creation of an irreducibly complex system, but I think Dr Behe himself put it best: 'I quite agree that my argument against evolution does not add up to a logical proof.'

The other widely used argument in favor of intelligent design is mathematical modeling. Again, there are two components. The first premise is that if something does not arise out of necessity or by chance, then it must have been designed by an intelligent being using a so-called 'specified complexity,' which basically means that if something resembles an entity that was derived independently by humans, for example, then it is specified in its complexity and therefore must also have been designed by an intelligent force. (Again, notice that the proponents of intelligent design do not utter the word 'God' and do not ascribe it to a specific religious agenda, because they know that this would destroy the sliver of legitimacy that they think they have built.) This complexity is demonstrated in mathematical formulae that describe random acts and deliberate strings of numbers. It is basically a challenge to the idea that if you generate a random set of letters for long enough a period of time you are very unlikely to come up with the text of *Moby Dick*. Those who push this idea, like those on the biology side of intelligent design, tend to present them in books and lectures to non-scientists.

The crowds who buy into this argument are dazzled by the math, mainly because they don't understand it. But here's the rub: you don't have to understand the math to understand that the basic premise is flawed. Nothing in life has any kind of specified complexity. That the human eye is similar to a camera lens does not mean that it was designed like a camera lens.

Evolution involves slow, subtle, random changes that are kept if they aid survival, are eliminated if they hinder survival and are left to drift if they have no effect in the animal. In addition, the drifting variations in DNA can at some point become relevant and be selected for or against eons after they first arose. If a subtle change helps one animal survive better than another, then over a period of generations, the one with the advantage will dominate and the other will disappear. Such is the way with whole organisms and such is the way with DNA. There is no set endpoint, no aim and therefore no design. A sophisticated part of the system can just as easily be selected against as it is selected for. Whales, for example descended from land mammals that evolved and adapted to live in the water. This is evidenced in genomic and anatomical studies. Are we to believe that the gain and then loss of the ability to live on land was part of a grand scheme to make complex structures designed for land and then let them slip away? The well-established goal of organisms is to survive and reproduce. If the development or removal of a complex structure will help that task, the organism evolves. The camera and many other human inventions actually mimic nature, not vice versa, because inventors often understand or come to the conclusion that the natural world has evolved efficient machines that are very good at what they do.

The other mathematical approach that has been used to argue for intelligent design is the cooption of a series of theorems called No Free Lunch. Basically, the math formulae associated with No Free Lunch have been used by intelligent design proponents to say that if you compare Darwinian adaptation through random mutation to purely random processes without selection that they are equal when it comes to arriving at an end point. That would lead one to conclude that natural selection has nothing to do with adaptation to the environment: the process or end point has to be predetermined in order to get there. However, it has been clearly shown that the model cannot account for animals evolving along with other living organisms. For example, every living organism on Earth comes into contact with other organisms and they react to each other. This is the foundation of evolution. Animals will adapt to predators, prey, vegetative food sources and pathogens. Like the 'specified design' idea, it falls apart when you actually consider how the living world works and cannot account for the variety of interactions in the living world that evolution can.

One of the scientists who originally came up with the No Free Lunch mathematical theorems has denounced their use in explaining intelligent design as flawed. And the man who is the biggest proponent of mathematical explanations of intelligent design, William Dembski, a fellow with the Discovery Institute, has been squashed like a bug. Once claiming that the theorems 'dash any hope of generating specified complexity via evolutionary algorithms,' Dembski now cowers: 'I never argued that the N.F.L. theorems provide a direct refutation of Darwinism.' His fatal mistake was arrogantly stepping into the arena with well-informed math-savvy evolutionary biologists and directly challenging a fundamental tenet of life with no real evidence and a poorly conceived mathematical argument. Intelligent design proponents will rarely do this because they are ill prepared and generally get turned into someone's intellectual bitch.

That intelligent design is laughable has not protected some children from being taught it in high school biology classes as a legitimate scientific alternative to evolution. One high-profile instance has been in the Dover, Pennsylvania school district where the school board laid down a new policy in October 2004, stating that the 'students will be made aware of gaps/problems in Darwin's theory and of other theories of evolution including, but not limited to, intelligent design.' In an interesting twist, all eight members of the Dover, Pennsylvania school board who had supported the introduction of intelligent design into the school curriculum were voted out of office in November 2005. Former Senator Santorum of Pennsylvania (I do like the ring of that) was quoted as saying 'intelligent design is a legitimate scientific theory that should be taught in science classes.' He was not the only intelligent design proponent to lose their job in the 2006 election. Several members of the Kansas Board of Education were ousted too and intelligent design will not be officially taught in Kansas. But this victory is not a reason to celebrate. It represents a close call that should make us all shiver.

A pressing matter

We know the public are not getting their science knowledge from the classroom, so how exactly are they making decisions and forming opinions on health and science issues? Simple: they are getting it from their religious and political leaders to some extent, but the majority get their information from the press, and even more specifically from television. It is easy to criticize television for reporting watered down science that builds up false expectations for viewers and generally over-hyping scientific achievement, but their most egregious fault is the rate at which they just plain get it wrong.

The written news is generally far better. But here too there are serious issues with their coverage. It should be made clear that the fault does not fall solely on the heads of journalists. Most people writing about health and science do not have any advanced education in the topic they are covering. Many will do their best to get the story right and there are a couple of mainstream newspapers that do a pretty good job. However, a report in the *Canadian Medical Association Journal* in 2004 tried to make the point that the press has done a good job on science stories. The researchers who did the analysis examined the accuracy of 627 newspaper articles published in the US, UK and Canada which were based on 111 peer-reviewed research papers. A close look at their numbers shows that a significant proportion of the time, journalists made significant factual errors or exaggerated the implications of the findings, even when they had the original research paper in hand. They found that 82% of the newspaper articles did not contain any significant errors and that 63% did not report any claims that were exaggerated above what the researchers themselves had claimed. Extrapolating, that means that 18% of the articles that appeared in major newspapers had significant scientific or technical errors and that 37% of them contained significant exaggerations of the scientific claims. It was interesting to note that the newspaper stories that reported on their report card actually spun the story to give themselves a pat on the back, claiming that the report confirmed that newspapers get science stories right most of the time. The question is: what other area of reporting would accept this rate of error (reporting on the love lives of celebrities excluded)? If this were politics, conservatives would cry a great liberal conspiracy. Whether this is the result of the public being science-stupid or the cause of some of the misconceptions about science seems unclear. Both are probably true to some degree.

Unfortunately, accuracy in reporting science and health stories is not the only problem. Which stories actually get covered is even more disturbing. Science stories are almost always based on new research findings, which on the surface makes sense, but producers and editors who have no working knowledge of science usually cannot make heads or tails of the actual science, and thus rely on press releases and catchy titles to determine what they cover. It's like buying clothes by window shopping, and never entering the store or actually trying on the garments. Thus there is a great tendency for them to cover stories on dieting to slim your big ol' fat ass, or how to protect yourself from the Sun. They correctly assume a low education level and almost always put some cutesy spin on how science will help you. You rarely hear of reporting of great advances in biology that are unrelated to human health, and the most common form of exaggeration in journalistic coverage of science is of the implications for health. The media have decided that the public don't care about or are unable to comprehend scientific advance, so rather than raise the public consciousness and report advances (which are news), they rally around the sensational stories in science.

This brings up an important point: what is news? For science reporting it would seem that the job of the press has been reduced to what the public wants to hear rather than reporting new discovery. Relevance is usually judged by what the public finds interesting rather than what they ought to be aware of or what is in fact strong scientific discovery. New findings that don't rise to the level of importance in the scientific community are often covered vigorously by the press because they do not have the ability to tell good science from bad science. This means that in addition to misreporting and exaggerating, the press misses the boat on discovery and reports minor advances as groundbreaking. There are several excellent science journalism schools that are trying hard to change the equation, but they are outnumbered or overcome by editors and producers that prefer the shiny happy face of a fluff story to hard journalism. This must change if the public is to grasp fully the true nature of scientific advance. The point that discovery is important, whether it is directly relevant to disease or lifestyle, seems completely lost.

True investigative journalism on science or health in newspapers and on television news is extremely rare. Journalists often just don't have a true connection to the scientific or medical communities to be able to probe into research and pull stories. This is radically different from many other areas of reporting, such as politics, where experience and relationships with the people they are covering are valued. The arrogance and sense of self-importance within the television news industry is absolutely nauseating in the face of ever lowering standards, piss-poor ability to find anything worth covering and shallow coverage of important issues. The role of the news industry has shifted from reporting what is happening in the world to entertaining the audience. You are more likely to see coverage of house fires and the abduction of children (when those children are white and attractive).

Blaming corporate America, news agencies and journalists would be giving only half the picture. Science journals and press departments within institutions as well as scientists are also to blame. The press releases and scienctific papers themselves are often the origin of exaggerated claims about science achievement. The goal of press releases is to inform the press of new research and to entice them to cover stories. So they often contain spin that inflates the implications of the work. On top of that, most scientists don't often get approached by the press, and are quite flattered when their work is recognized in the media. While throughout an interview a scientist may be very careful to point out the limitations of the results, one spicy quote is often all an eager journalist is looking for to advance the story.

Some news agencies have off-air science and health consultants that help them to evaluate the strength of scientific stories and suggest stories that will have implications for the public or that they feel are important enough for the public to be aware of. On the whole the relationship works very well, and results in stronger coverage. If more news agencies took such an approach the level of coverage would improve dramatically, which would be a great service to the public, who are in great need of quality information on science and medicine.

Making it

A final note on the danger of a poor science education system: most people would be shocked to learn that the people in Congress that create science and healthcare law usually have no background in science or healthcare. I'm not talking about the actual elected congressmen; I am talking about their staff. Now, for most issues relating to health and science, one does not need to know anything about treating people. But for many issues, the only thing driving the law is advocacy groups that reach and pull the heart strings of the staff and congressmen. Since they rarely have any expertise and since they often do not get advice from objective scientists and doctors, much of the healthcare law that is proposed is misguided. Lobbying on the whole can be a very good thing for ignored diseases, but the one saving grace is that NIH money is not earmarked for specific research objectives. In other words, scientists set the agenda based on what they think they can accomplish. There are some in Congress who feel that they should be dictating where research should go, but with no scientific or health expertise to back their opinions this is dangerous.

The collective failure of the US public school system has occurred under the harsh light of a nation continually exposed to its own self-deception and hypocrisy. It is widely recognized throughout the world that our preoccupation with self-aggrandizing rhetoric and patriotic flag waving has left us ill prepared to confront our own weaknesses. Our leaders spend more time posturing over trivial matters than actually addressing the fact that our own rigidity and aversion to risk has resulted in a bureaucratic infrastructure that is largely ineffective. Unfortunately, the education system is a prime example of this, and parents will continue to contribute to the problem as the trend towards a preoccupation with educational status continues at the expense of a focus on, or even interest in, what their children are learning. This vicious cycle will continue as poorly executed policies like No Child Left Behind continue to erode the learning atmosphere of schools by forcing teachers to teach children to be better test takers than thinkers. Science education will continue to falter unless there is a radical change in the value that US citizens place on education as a whole. This must start at the top, with our leaders understanding that the enormous opportunities that lie ahead in science and medicine will not be taken by American youth, but by foreign students who value the opportunities before them. The saving grace will be that when the situation reaches a catastrophic phase, we'll have enough lawyers to argue over whose fault it is.

and another thing: Scientists and Engineers for America

Anti-science policies and views in the US have gone unchecked for too long. The idea that ignorance of science would be caught in the standard societal checks and balances that protect us from extremist views has been proven wrong. Without serious action, the retreat of reason will get worse, science education will get worse, funding of science will get worse and our nation will be in serious trouble as the economic and social ramifications of this decay become more apparent. It is high time that researchers, teachers and those who believe in the fundamental value of science come down from the ivory tower, step up and make themselves heard. We must actively engage in discussion of the benefits of discovery and aggressively debunk the myths about science commonly held in society.

In September 2006 I had the pleasure of being involved in the formation of a new political organization, Scientists and

Engineers for America. SEA is dedicated to electing public officials who respect evidence and understand the importance of science to public policy. We grew to over 8000 members in less than a month and were active in several key states for the 2006 elections. We have a prestigious Board of Advisors that includes over a dozen Nobel Laureates and will be setting up student chapters around the country. For more information please visit www.SEforA.org.

Grass roots efforts can make all the difference if people are willing to participate. I am talking about minutes of time to email a letter and hours of time to write an editorial for a local newspaper or address a group in public. That is not a big commitment to increasing support for the future of science in the US – and hence the world. Even if you just join SEA to keep up with what is happening in the current war against the politicalization of science, it is more than you are likely doing now.

In the first printing of this book I proposed a National Science Day on 2 October every year. That idea is still alive and well. Think about it: if scientists take the time to frankly discuss the value of science with students, businesses and faith leaders, people might start to understand how far we have strayed from a time when the efforts of scientists were held in the highest regard. If enough of us take part in this dialogue it will send a message to the nation that cutting funding for science is not right, that the misrepresentation of scientific results by our leaders is not right, that the lying about and manipulating of science to serve a religious, economic or political agenda is wrong, and that the ignorance of science that has spread through the US is a travesty.

Why wait? Write to your congressmen now asking them to introduce and pass a resolution recognizing 2 October as National Science Day. It doesn't matter if you voted for them or not: if you live in their district and have a problem with the way things are going, tell them. Just write a brief email explaining what your issue is.

Hell, what is there to lose? In the past three years, Congress has introduced resolutions designating many days, including National Marina Day, National Day of the American Cowboy, National Tartan Day, National Airborne Day, National Child Care Worthy Wage Day, National Asbestos Awareness Day, National Kindergarten Recognition Day, National Dyspraxia Awareness Day, National Hunger Awareness Day, National Time Out Day, National 'It's Academic' Television Quiz Show Day, National Good Neighbor Day, National Native American Veterans Day, Lights On Afterschool! Day, and National Mammography Day.

No matter what you do, do something. If you take anything from this book, make it the fact that you can make a difference. Get involved and stay informed of what is going on in the US. It does not take a lot of time or effort to do your part. Please visit www.SexDrugsandDNA.com or www.SEforA.org for more information on what you can do to get involved.

essential reading and resources

It is paramount that the public be able to find accurate sources of information on science, health and the associated policy. Since the administration has taken a radical turn away from providing the public with reliable information, I have compiled a short list of essential resources.

- *The Republican War on Science*
  Journalist Chris Mooney does a stellar job of documenting the efforts of the conservative movement, backed by religious fundamentalists and a corporate machine that use Republican politicians as a battering ram to pass policies that suit their agenda. Read this book.

- *Flying Blind: The Rise, Fall, and Possible Resurrection of Science Policy Advice in the United States*
  This report by the Federation of American Scientists details the history and current state of science advice that the government is receiving. Anyone who is wondering why Congress and the President seem so wayward in their

approach to health and science should read it. You can view it for free at www.fas.org/static/pubs.jsp.

The Federation is also a very useful resource for information on a variety of other important issues.

■ The National Academies of Science
This last vestige of high-quality scientific assessment in Washington has not been corrupted by the politics. If you want information on the state of any scientific or health issue, turn to them first. Their reports can be read for free at www.nas.edu.

■ The Trust for America's Health
This non-profit organization is an excellent resource for accurate information on health and health policy. Their assessment of health policy is spot on and their recommendations are accessible and sensible. They are a strong voice of reason in Washington; www.healthyamericans.org.

■ The Kaiser Family Foundation
When it comes to information on healthcare, you cannot top the information available through this non-profit foundation. Their assessments of our healthcare system are truly remarkable; www.kff.org.

■ The National Center for Science Education:
www.ncseweb.org

and

■ The National Science Teachers Association
www.nsta.org
The final word on science education comes from the teachers. Listen to them and learn.

Index

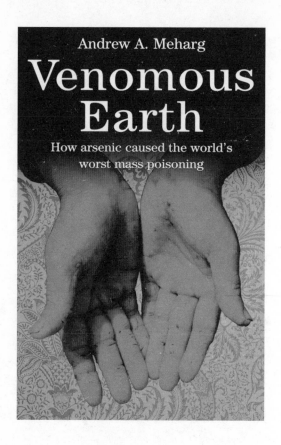

VENOMOUS EARTH
HOW ARSENIC CAUSED THE WORLD'S WORST MASS POISONING
by Andrew Meharg
MACMILLAN; ISBN: 1–4039–4499–7 £16.99/US$29.95;
HARDCOVER

"Meharg is good on the technological and political challenges of testing water. He is terrific on the wider history of arsenic, in alchemy, industry and interior decorating." *Guardian*

"Meharg tells the lively and cautionary story of arsenic's misuse over the centuries." *Newsweek International*

order now from www.macmillanscience.com

THE WHOLE STORY

ALTERNATIVE
MEDICINE
ON TRIAL?

TOBY MURCOTT

THE WHOLE STORY
ALTERNATIVE MEDICINE ON TRIAL?
by Toby Murcott
MACMILLAN; ISBN: 1–4039–4500–4; £16.99/$24.95; HARDCOVER
ISBN: 978–0–230–00753–6; £8.99; PAPERBACK

"Essential reading. A balanced, sympathetic and long overdue look at the relationship between science and complementary medicine." *Focus Magazine*

"This book should be prescribed for bigots on both sides, to be taken thoughtfully, all the way to the last page." *Guardian*

order now from www.macmillanscience.com

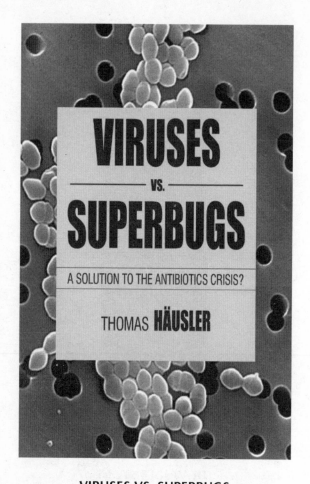

VIRUSES VS. SUPERBUGS
A SOLUTION TO THE ANTIBIOTICS CRISIS?
by Thomas Häusler
MACMILLAN; ISBN 978–1–4039–8764–8; $24.95/£16.99;
HARDCOVER

'Valuable reading, both for specialists and for interested general readers.' *JAMA*

'All the ingredients of a John Le Carré spy novel: fascinating.' *EMBO Reports*

order now from www.macmillanscience.com